The Split-Second Integral

—

The Derivation of the Instant

—

Object 4D Rotation in the Space-Time/Manifold

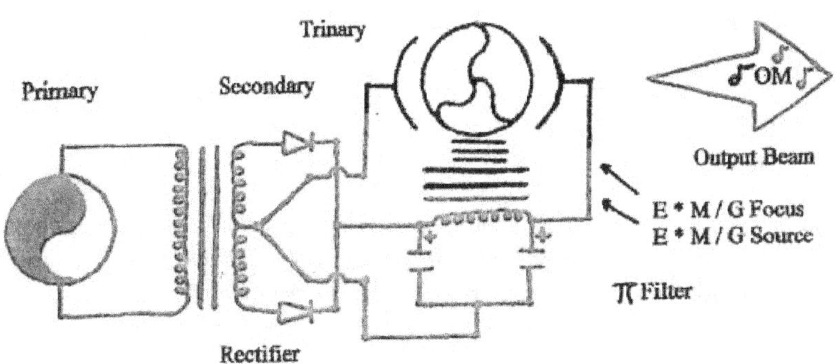

by William Holland

Dorrance Publishing Co
585 Alpha Drive
Suite 103
Pittsburgh, PA 15238
Visit our website at *www.dorrancebookstore.com*

ISBN: 979-8-88729-350-9
eISBN: 979-8-88729-850-4

INTRODUCTION

I have always loved science fiction, the stories have been great. What I enjoy most is the physics of ultralight movement and time travel. Yet to this day I cannot find a reasonable book that accurately describes the working physics of the space-time wedge or an ultralight scalar field drive motor. Keeping this in mind, I decided to create a book based upon the actual physics and mathematics of space-time movement, a book that accurately describes how to move an object to a new, chosen space-time frame.

The text starts with a very basic history of physics, including the necessary mathematics to understand the movement of an object in space-time. It progresses through the transition of three-dimensional logic to the time-based four-dimensional logic while describing the Hypercube and four-dimensional movement. I explain the mathematics of the Split-Second Integral and its Trinary Field Generator Array, then I explain the mathematics of the Derivation of the Instant and how to rotate an object in all four dimensions of the space-time/manifold. I add some interesting diagrams explaining the control of the necessary scalar fields used to propel an object to anywhere and anywhen that is desired.

My final chapter deals with Cosmic Energy and the creative energies of the mind's life force.

We can split the second and open an entire new universe! Superliminal Velocities and Time Travel would be at our command; communication would become instantaneous, computation speeds unlimited. We could derive any moment anywhere in space or time and we could move a chosen object through the four dimensions of the space-time/manifold to that new location.

This collection is the basic history and the mathematics of those discoveries that will explain the Split-Second Integral, the Derivation of the Instant, and the controlled rotation of an object in the four-dimensional space-time/manifold. After I fill in the missing pieces, I will generate the necessary plan to develop the Trinary Field Generator Array that will allow us to split the fabric of space and time, permitting travel through the higher dimensions of Hyperspace.

Matter (the atom) has already been split; with these formulas, we will be able to split Time itself.

With the discovery of the Quantum Physics a new door has been opened, a door that had previously been thought impossible to open using the Classical Physics.

Scientists have actually shown that faster-than-light communication and reversed time motion is indeed a reality. This book will reveal the exact format used by scientists to create the necessary scalar fields to generate ultralight and reversed time motion.

When I first wrote this book, I was told I would not be able to publish it due to the sensitivity of its content. With the Freedom of Information Act, the rules have changed, allowing the publication of this type of material. I have done a lot of research into some of the more difficult aspects of quantum physics so as to make it a lot easier for the reader to understand the basic concepts of the actual movements in the space-time continuum. With this platform, I progress to the Split-Second Integral, the Derivation of the Instant, and the four-dimensional rotation of an object in the space-time/manifold. These are the formulas that will indeed make all of this a reality.

The final diagrams are of the Trinary Field Generator Array, which is all based upon proven data. The Trinary itself generates the necessary scalar waves, which are focused and in turn excite the necessary ultralight and reversed energy fields.

I hope that this book will help to explain much of the declassified material that is now available today, and I hope that you, the reader, in pursuit of knowledge, find this book as interesting reading as I have had in writing it.

CONTENTS

A Brief History of
Classical and Quantum Physics

The Classical Physics of Isaac Newton was completely sound and acceptable up until about the 1900s. The Mathematical Tools of the Classical Physics worked very effectively at the lower speeds that did not include relativistic velocities or at velocities approaching that of light.

Problems developed with the measurements of time and distance as particle motion at relativistic velocities were being measured.

It seems that [time] x [distance] was not equal to [distance] x [time] at velocities near that of light, as it should be. The problem was solved with the addition of a new mathematical element called "Imaginary," expressed as (i). This value (i) showed that at speeds approaching that of light, Extra Dimensional Motion takes place. Another way of looking at it is that Higher Dimensions occur. When the Imaginary value (i) is squared $(i)^2$, it becomes a negative number (-1). It has in fact moved "Forward Through Negative Time" or "Backward in Time." The value (i) is infinitesimally small, but when [time] x [distance] is multiplied by (i), it becomes equal to [distance] x [time].

So [time] x [distance] x (i) = [distance] x [time] at relativistic velocities.

This leads to the idea that time itself is not rigid but is a more fluidic motion.

Quantum Physics discovered that time is fluidic and has Imaginary movements. These Imaginary movements involve Reversed Time or Reversed Motion at relativistic velocities. Quantum Physics also has Energy transmitted in packets and not the steps or levels of Classical Physics, proven by Einstein's paper on the photo-electric effect.

As theory progressed, four-dimensional logic came into being. Height, width, and depth were the three dimensions x, y, and z drawn at 90-degree angles to each other as the first basic coordinate system. Minkowski added a fourth 90-degree angle or 90-degree rotation by adding a Time Axis or Time Dimension. Any object in space-time now had a four-dimensional coordinate (x, y, z, t). This system is known as Minkowski Space.

At about this time, Einstein published two important papers. His introductory work concerned the Electromagnetic Field and the Gravitational Field with its effect upon the Inertia of Mass. Einstein's Special Theory of Relativity briefly states that light, in a vacuum, is the upper speed limit of this universe. No object can travel faster that light without reversing the direction of time. Using vectors, Lorentz transforms, and measuring rods, Einstein proved that matter approaches infinite mass or weight as it approaches the velocity of light. The object can never achieve the velocity of light or exceed it because it would become as heavy as the entire weight of the universe itself, which is, of course, impossible. Furthermore, if an object is going at near light speed in one direction and a second object is going at near light speed in the opposite direction, the two can only be separating at light velocity, not near twice light velocity, as their mutual gravitational attraction forbids either from separating from the other at a velocity greater than that of light, in a vacuum, in this universe.

Einstein's General Theory of Relativity states that Matter itself bends the fabric of space-time, forming a dip, a pool, or a conic that is very much like a ball rolling on a sheet of fabric. His equations prove that Matter is three-fourths energy of the Electromagnetic Field and one-fourth energy of the Gravitational Field. He also proved Mass becomes Energy at Light Velocity and its compliment Energy can become Mass.

About 1930 to 1940, physicists were looking for an explanation of the strange behavior of the electron. They used the quantum physics imaginary tool (i) to discover the electrons mirror-matter or anti-matter counterpart, the positron.

The positron is a particle exactly like the electron except that it is moving "Forward Through Negative Time" or "Backward in Time." The positron's energy field is Negative (-1) and Real. The discovery of Mirror-Matter or Anti-Matter led to the reality of Extra-Dimensions. The Universe itself has these Extra-Dimensions because it contains forward and reversed time movements.

These Dimensions can be listed as:

Height (Future)	Width (Future)	Depth (Future)
Height (Present)	Width (Present)	Depth (Present)
Height (Past)	Width (Past)	Depth (Past)

Time (Future)   Time (Past)

Time in the present is instantaneous and can be considered Nondimensional, as the Object Mass is actually Energy (three-fourths Electromagnetism and one-fourth Gravitational) at the instantaneous light energy equivalency in equilibrium. Time of the Future and the Pasts Extra-Dimensions are there but are infinitely smaller than the Object itself. As the Future becomes the Present, the Present becomes the Past. To Time Travel is to change the location of the Objects Present on the Object's path or World Line.

Objects Present (x, y, z, t)

Future (+x, +y, +z, +t) * NOW * Past (-x, -y, -z, -t)

[Object's path or World Line]

Some very interesting results have been discovered using the new Quantum Physics. Physicists have discovered that particles of matter exhibit two basic characteristics. The first is that a particle of matter appears to exist as a solid object. The second is that the same particle appears to exist as a wavelike potential. This was discovered when particles were found to be passing through an impenetrable barrier. In Classical Physics, the particles aimed at an impenetrable barrier could not possibly pass through the barrier. Physicists concluded that the Wavelike Duality of the particle "Tunneled" through the barrier. This can best be explained as some of the particles Rotated through the Fourth Dimension of Time and reappeared on the opposite side of the barrier (Fig. 1).

Recently, physicists in Germany have assembled a working InfraRed Photon Communicator that focuses photons through a solid barrier, resulting in communications that are in excess of light speed. The first results were just over light speed, and the later results were many times the velocity of light, with no signal degradation.

The Double Slit Experiment

This experiment is the famous proof that we exist in a Quantum Universe, not a Classical one.

We are given that electrons and photons have been proven to be particles, in nature, by the mathematics of Einstein.

Electrons or photons are emitted slowly and at random angles from a point source through a dual-slit array to a viewing screen or photographic plate (Fig. 2).

Classical Physics would be proven true if just two separate lines appeared (one behind each slit), or, eventually, the entire area would be filled with no lines or fringe patterns but just an even textured exposure. Particles would be just that, particles with no wave characteristics or wave duality.

This is not the case, however, as fringe patterns or interferometry waves form (waves of dark lines and bright lines). This is the proof that the wave-particle duality of quantum physics is the actual true nature of this universe. The particles themselves exhibit the characteristics of waves as they pass through the dual-slit array. Even though one particle at a time is emitted through either of the two slits, a wave pattern forms. In the case of the photon, the interference fringes form even as one photon at a time is emitted. In this case, only the photon can interfere with itself! The cause may be another dimension blocking it, the actual idea being that since particles are emitted only one at a time, there is nothing to interfere with them, yet they develop fringe or interference patterns. They must be being interfered with by some extraneous or multi-dimensional presence. That presence may be the same photon on a different space-time path blocking the original path itself and causing interference waves to appear as it interferes with its own presence.

Penetration of an Impenetrable Barrier

Classical Physics: An Object cannot pass through the impenetrable barrier.

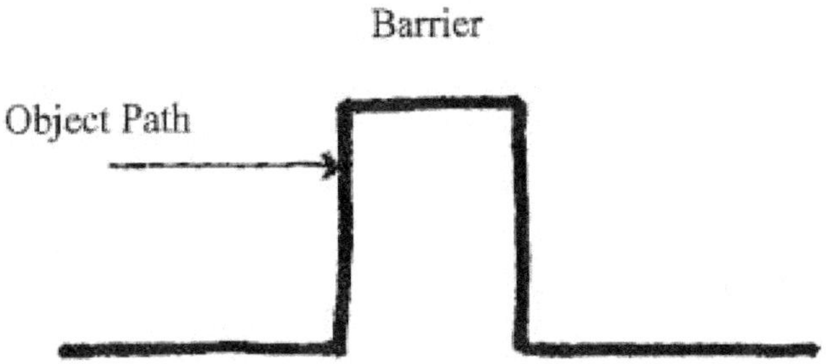

Quantum Physics: The Object "Tunnels" through the impenetrable barrier by traveling through "Reversed Time in Imaginary Space."
(The dotted lines path)

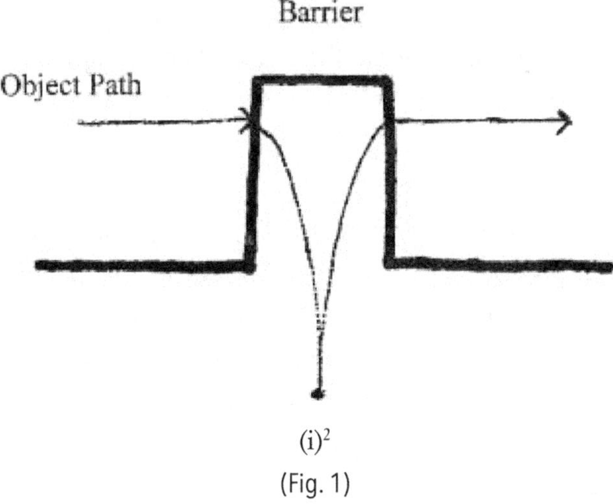

$$(i)^2$$

(Fig. 1)

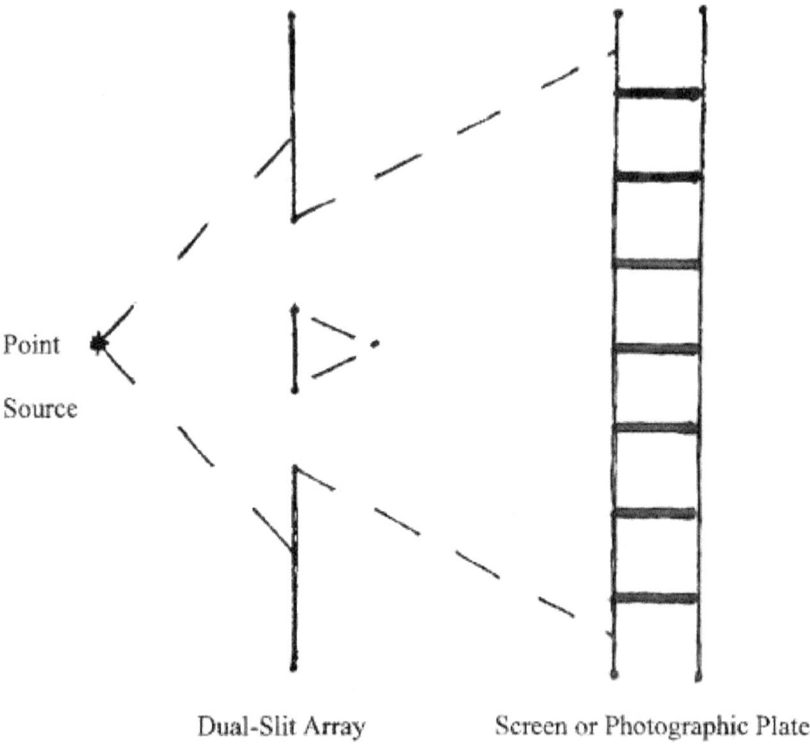

Electrons or Photons are emitted from a Point Source through a DualSlit Array. Fringe or Interferometry Patterns form, proving the Quantum Physics. Classical Physics would be the conclusion if only two bands of light formed (one behind each slit).

(Fig. 2)

Several scientists have postulated the existence of a multidimensional universe based upon these given results. They postulate that photons (as light waves) can be simultaneously connected to two or more separate space time frames or events at the same time. An expansion of this idea is that an object can be quantumly or infinitely connected to two space time frames or events, resulting in space-time travel to anywhere or anywhen.

It has been proven that a multi-universe or multi-verse is possible using quantum mechanics. In this view, there are an infinite number of universes or space-times existing side by side, with gravity (the weak force in open loops) being the glue between each and light or photons (the strong force in closed loops) able to just touch two space-time events for each moment or instant in time.

With this in mind, scientists have concluded that this universe is just part of the Multi-Universe. This has been compared to a book, with our universe being a page of the entire book. As explained, this Universe has Matter that is composed of Electromagnetism (the Strong Force) in Closed Loops, which is in combination with Gravity (the Weak Force) in Open Loops.

The Matter is Closed Looped and stays in this Universe, where it is weakly affected by Gravity, which is Open Looped, and so is in partial contact with its Neighboring Universes. Gravity is the bonding agent between universes. Its effect is exactly like the Quantum Tunneling that was explained earlier.

It has been postulated that Time Travel may simply be Quantum Tunneling to anyone of the Parallel Universes.

Interesting Mathematics, Facts, and Definitations

To make things a bit easier to understand, I have decided to include a brief mathematics review on just the necessary Physics, Facts, and Definitions that are required to explain an object's movement in space and time.

Space itself is a vacuum; it is timeless and infinite. It is as large as the universe and as small as nothing.

All objects are composed of matter and are in this vacuum of space.

Matter is composed of Fields of Electromagnetism (the Strong Force) and is weakly affected by Gravity (the Weak Force).

This effect is described as the Object Mass or Weight.

All matter tends to remain at rest or in uniform motion (unchanged motion) unless acted upon by an external force (energy). This is called Inertia. Gravitational Reactance with Mass is Inertia and is described mathematically by Einstein as $\sqrt{-g} = 1$. The path of movement that an object takes in space-time is mathematically expressed as the xct axis. The (x) is the object's (x, y, z) coordinate in space; the (c) is its inertial reference to the velocity of light; and the (t) is its time reference. The xct axis is used as a shorthand notation for computing the object's inertial path through the higher geometries of the Multi-Dimensional Riemann Space.

The Objects Momentum is its property of movement that determines the length of time required to bring it to rest when under the action of a constant force or movement.

A Vector is a quantity that has magnitude, direction, and is commonly represented by a directed line segment whose length represents the magnitude and whose orientation in space represents the direction.

An objects movement in space-time is represented by a Vector. Physicists use "u" and "v" in many of their quantum equations as their standard vector notation.

Vectors can be added together in the Multi-Dimensional Riemann Space to change the original xct inertial path of the object so that it arrives at its new chosen location by using Lorentz Transformations, Matrices, and Determinates.

The Matrix is a rectangular array of the objects mathematical elements expressed as coefficients of simultaneous linear equations. An example is the object's (x, y, z, t) coordinate location or anchor point in time, with the second matrix being the desired or chosen space-time endpoint.

The Determinate is the sum of all the products of two or more matrices, forming a single linear equation, such as an object's xct inertial axis path from its anchor point matrix to its endpoint matrix.

Lorentz Transformations of Motion are the mathematical equations that deal with the application of Energy (Impulse) to the object's xct axis or Inertial Path, to change its Energy Level in the space-time/manifold.

Hermitean Equations of Energy for Finite Energy Perturbations are the balanced equations for the flow path of Energy through the Nucleus of Atomic Matter. Infinitesimal perturbations of Field Energy are used in Quantum Physics to measure infinitesimal changes in the Spherical Harmonic Nuclear Superstructure of an object's mass. This process shifts Energy Flow through the Atomic Nucleus to achieve a chosen result. In Chemistry, equations are setup so (A + B) = (C + D). In Physics, the Hermitean Energy Equations are setup so (A+ B)- (C + D) = 0.

A Tensor is defined as a vector with a double row of indices that have invariant properties under transformation of the coordinate system. A Tensor generates a tension, stress, or curvature in the space-time continuum. I am interested in the Gravity Tensors that describe the effects of gravity waves upon an object's inertial or xct axis. The gravity tensor can best be described as "Depth" of an object in the fabric of space-time. The gravity wave is like a figure eight, with one loop around "NOW," then instantly one loop around

"THEN." The reaction of a gravity wave starts with the object at "NOW," or Tensor (g 0,0), and as it passes through the Object's Nuclear Superstructure, it creates a Spherical Harmonic Resonance with the Object's Electromagnetic Field and moves the Object's Inertial Moment to "THEN," or Tensor (g 1,1). The reaction of a Gravity Field with the Object's Electromagnetic Field starts at Tensor (g 0,1) and concludes with Tensor (g 1,0), so that the object arrives at "THEN." In brief, the Gravity Tensor Path through the Object's Nuclear Superstructure is (g 0,0) ~ (g 0,1) ~ (g 1,0) ~ (g 1,1), with the Quantum Resonance Interval (g 0,1) ~ (g 1,0) being the Moment of Inertial Energy Transfer as the Object's Electromagnetic Field goes into Spherical Harmonic Resonance with the applied Gravity Field.

The Gravity Field has an Infinitesimal Perturbation (penetration) for each Instant of Time upon the Electromagnetic Field of the Object, but the total summation of the Gravity Field is totally absorbed by the Electromagnetic Field at the Object's Spherical Harmonic Resonance (the point where all three x, y, and z axises of the object have their maximum surface area or contours exposed). The angle from (g 0,0) through (g 1,1) is called the Pointeen Vector.

The (g 0,0) Membrane is Absolute Gravity, where the object's electrons are within the nucleus (0 ~ 0).

The (g 1,1) is Absolute Connectivity, where the object is Quantum Tunneling along the curvature or circumference of the universe. The object is in Synchronization or Heterodyned to the specific chosen Scalar Photon Field(s), one field from "NOW" and one field from "THEN."

The (g 0,0) Membrane ~ The (g 1,1) Universal Curvature or Circumference. See Fig. 3.

Consider the absolute vacuum of space-time; it has no future, present, or past unless there is an object within it. The object is matter or energy and has a surface or contour in space-time. The object started from nothing and is expanding with the universe to the infinite. At the first moment of its existence, it was on a surface called the (g 0,0) Membrane. It is then that its matter or energy became part of the space-time of this universe. Its gravitational field is connected with all other objects in this universe because each other object in this universe was originally connected to the same (g 0,0) Membrane. The (g 0,0) Membrane is in essence a Platform or Anchor Point for all of the objects in this universe's space-time continuum.

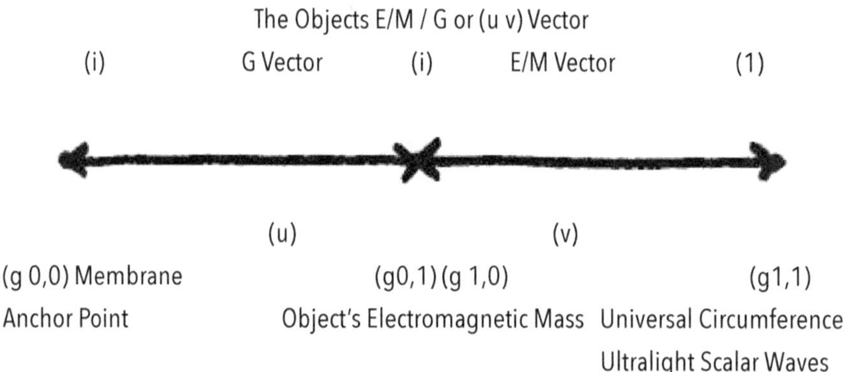

The G Vector (u) from (g 0,0) to (g 0,1) represents the Object's Gravitational Energy. It links the Anchor Point at the (g 0,0) Membrane, or Schwartz Field Radius (Event Horizon) to the Objects Nuclear Mass.

The E/M Vector (v) from (g 1,0) to (g 1,1) links the Object Electromagnetic Energy to the Scalar Photons at the Universal Circumference.

The entire energy Vector (u v) or (g 0,0) ~ (g 1,1) can be described as the Object's Tensor, with the starting and ending points being the Object's Mathematical Matrix Elements and the sum of these two matrices being its Determinate.

The entire energy wave that this describes is called the Object's Soliton Wave. It represents the object's xct inertial path from the origin, or Pointeen Vector, through its Nuclear Superstructure, computed by the Hermitean Equations, to the object's endpoint, which is its resulting mass along the Universal Circumference.

(Fig. 3)

An object can change its space-time location by altering its Electromagnetic and Gravitational Fields.

Lorentz Transformations are used to compute the necessary Energy Impulse that an object needs to move to anywhere or anywhen or both. This is done by the generation of an Energy Impulse that is tuned to the Object's Spherical Harmonic Resonance. The actual movements are vectors in the Multidimensional Riemann Space.

Changing the Electromagnetic Field of the object alters that object's (g 0,1) ~ (g 1,0) Gravity Tensor in space-time. The result is a change in the Inertial Motion of the object. The object moves along a new path. Controlling the change results in a controlled space-time relocation to a new, desired space-time location.

The (g 0,0) Membrane, or Schwartz Field Radius (Event Horizon), is the focus point of the Field Equations used to duplicate the conditions of the Event Horizon (zero - zero) space-time, and to travel through Hyperspace $(1)^3$ [Future] or Sub-Space $(i)^3$ [Past]. This is the equivalent of travel through Infinite Positive Time {Idea} and Height in Time or Reversed Time (Anti-Time) and Depth in Time or {Memory}. This is the point where matter is compressed to "zero-dimensions." The object has no Radiant Temperature or Radiant Output and its electron radius equations go from External Nucleon Radius (r) to Internal Nucleon Radius (R). The Nucleon Wave Dynamic has no Radiant Energy. It becomes inert.

Gravity itself loops back from the Universal Circumference (g 1,1) through infinite dimensions, returning from (g 1,1) to (g 0,0). It is now "Memory," or "What Was." When the object is reduced to this (g 0,0) point, it leaves its Signature or its Photonic Image on the Universal Circumference at "Now" or the "Present."

This is the record of its reaction with all of the other objects in this universe, stored as Light (Electromagnetism and Gravity). These records become "scalars of light," or "soliton waves of energy."

The object can use the Infinite Connectivity at the (g 0,0) Membrane to select a new path and arrive at a new space-time location by accessing the infinite (zero-point) energies that are on the Universal Circumference at (g 1,1). Light can "escape under" the (g 0,0) membrane. When this occurs, the "time axis" changes direction. If an object's xct axis, or its "Inertial Time Axis," is reversed, it will change direction in time.

An object's electrons can be compressed into the nucleus by photons (light). The photons carry the radiant energy away from the nucleus at spherical harmonic resonance. Electrons spontaneously emit photons as they lose energy and finally go to the (R) radius (within the nucleus) at (t 0,0) time and (g 0,0) gravity.

The object now rests on the (g 0,0) membrane, where it has no apparent radiant energy, and it is not actually in any space-time frame. It can move to anywhere or anywhen from this (g 0,0) membrane if the correct energy impulse is applied to its nuclear mass.

Energy Density is the amount of energy in a given volume of space-time. Negative Energy Density is energy with density less than zero in the vacuum of space-time itself. Negative Energy exists and is real. It is defined "Imaginary" and its time axis direction is from the Future to the Present or from the Present to the Past. A Negative Density Energy Field can result in Time Reversal.

Positive Energy: $(1)^3$ or $(+1)$, or "Creation Energy." Energy that is here and now and will be here in the Future. Examples include Tachyons or Ultra-light Velocity, Multidimensional Energy, Double Time Forward, Travel Through Positive Time and The Future. Positive Energy is referred to as "Height in Time" or "Idea."

Mathematically expressed as $E = MC^2 / (1)^3$.

Negative Energy: $(i)^3$ or $(1-)$ or "One Then Minus." Energy that is in the Past. Examples include Positrons or Reversed Time or Travel Through Negative Time. Energy that is "Less Than Zero," of the "Past" or "Memory." Referred to as "Depth in Time." Negative Energy might exist in a space that has been relativistically deformed around an ultra-compact mass. At this point, the Neutron Permeability Constant (npk) exceeds Unity, and Matter "Falls Out" of this universe into the past. Mathematically expressed as $E = MC^2 / (i)^3$.

Facts

Light is Electromagnetism (Maxwell).

Matter is three-fourths Energy of the Electromagnetic Field (Light) and one-fourth Energy of the Gravitational Field (Einstein).

The $(0 \sim 0)$ electron reaction. The nuclear reaction where the electron is photonically compressed into the atomic nucleus, leaving the Atomic Nucleus

Wave Dynamic with no Radiant Energy (temperature). At this time, there is no Electromagnetic Field outside of the nucleus, and the nucleus has zero dimensions and no sense of time. It is no longer in the Present space time frame, and it is in a state similar to suspended animation. It has in effect become a supersolid, existing on the (g 0,0) Membrane where time has no effect.

Imaginary Negative Time. The Time Axis is negative or reversed, and time flows from the Future to the Present or from the Present to the Past. Imaginary (i) does not exist in "Today." It exists in "Yesterday" and affects "Today" by changing "Yesterday." Imaginary Mass also has a negative number and is flowing along the time axis in a reversed direction.

Photon Torque has recently been proven real. Photons have an additional "Torque," or "Spin," as well as angular momentum (push). This increases "Connectivity." The Photon Field Torque is used to move an object through space as well as time.

Cherenkov Radiation is a photon stream produced as electrons travel "Faster than Light" through a medium such as water. The actual velocity is (.75 C). The blue light is compared to a sonic boom and is called a "Light-Boom." Cherenkov Radiation is typically found in the reaction chamber of a nuclear pile.

Zero-Point Energy is found in the vacuum of space. Many physicists believe the vacuum of space is really a "Hot Bed of Forces." Phantom particles flicker into existence and then disappear. Empty space itself seethes with vacuum fluctuations as vast amounts of energy burst forth, jiggling particles to and fro. Cumulatively, they can be very intense. There is enough energy in the vacuum of a single light bulb to boil all the seas on Earth. The zero-point energy of the vacuum might prevent atoms from collapsing. Electrons should radiate their energy as they circle in their orbits. They do not drop into the nucleus because they absorb enough energy from the vacuum fluctuations to make up the loss. The stability of matter itself depends on the zero-point energy sea. Zero-Point Energy can be extracted from the vacuum of space using the Casmir Effect. If you bring two smooth plates of metal extremely close together, they will seem to attract each other so strongly that they will be virtually welded together. If two physical bodies are relatively close, the first shields the second from zero-point energy coming from its direction, just as the second shields the first. Both objects continue to be pressured by the zero-point energy coming from all other directions. The two bodies are moved towards each other as the Casmir Effect, which is what Classical Physics defined as gravity. This stress or tensor in space-time is a source of power. See definitions "Casmir Effect."

Quintessence is one of the most recent discoveries that has come to light. Quintessence is the real form of "Dark Energy" to account for two-thirds of this universe's energy density. Scientists are spending some serious time to prove this dark energy is here. They have shown that galaxy clusters, the largest objects in the universe, must contain only 5 percent part baryonic matter (the stuff we are made of), and approximately 30 pecent dark matter mixed with approximately 65 percent of this quintessence energy (dark energy) for everything to act the way that it does. Their proof is that distant supernovae are dimmer than they should be. This means that the universe is expanding faster than it should be. Dark energy is its cause. If the gravitational force of quintessence is repulsive, it would account for this acceleration. Dark energy resists the gravitational pull of the galaxies. One idea is that dark energy is really vacuum energy, the energy of empty space. All quantum fields possess a finite amount of zero-point vacuum energy (as a result of the uncertainty principle). Quintessence interacts with matter in a way that affects the forces between particles. If the quintessence field is varying temporally (in time) or spastically (in space), it will cause the forces between particles to change as well. This may lead to quantum gravity and inertial time control.

Cold Fusion is a recent idea that could be a great power source. Fusion itself is a nuclear reaction that produces tremendous amounts of energy by combining hydrogen plasma (hydrogen atoms without electrons), into helium nuclei at temperatures over ten million degrees Fahrenheit. This is the same reaction that fuels Earth's sun. The problems are numerous. The heat of fusion must be generated and contained. An alternate method is the Cold Fusion Cell. The cell uses a palladium cathode to hold the atoms of heavy hydrogen (deuterium) formed after the electrolysis of heavy water into deuterium and oxygen. A small control current is run through the cell using platinum as an anode and the deuterium saturated palladium as the cathode. The cell produces heat (power) as the deuterium nuclei fuse to helium while in the palladium matrix. Claims are 400 percent power gains.

The Caduceus Wound Coil and Scalar Waves

The first Caduceus Wound Coil was insulated copper wound in a double-helix about a ferrite core only three times, one for each plane (xy, xz, yz). It was developed by the Air Force around the 1950s when they were studying Electro-Magnetic Propulsion Systems. It was the only coil array found that produced Unexpected results when powered by RF (Radio Frequency) energy. When powered by pulsed bursts of microwave frequencies, these coils lift themselves and do periodic hops off the ground plane. This is attributed to the field effect produced by the unique coil windings. The three helix coil windings cancel the normal perpendicular magnetic fields generated by the standard electro-magnetic coil as they "dead end" the fields in each of the three planes (xy, xz, yz), or one per loop, and force a Scalar (Ultralight) Wave to form at the ends of the ferrite core.

The fields are Quantum Tunneling through the core. This magnetic field is parallel to the ends of the coil and is actually in excess of lightspeed forming Scalar (Phase) Waves. When a whip is cracked, its tip momentarily breaks the sound barrier, forming a "snap" as the air rushes back together. When an electron is "Dead-Ended," it momentarily turns opposite as its inertia reverses. This reversal in inertia produces a field effect that reverses the direction of gravity and time. It becomes a positron or low energy (microwave) photon.

This reaction can cascade with dark matter and dark energy in a multi-dimensional reaction to generate a Field Scalar (Phase Wave) that is "Faster than Light" and reverses the direction of time as well as that of gravity.

A Scalar Potential, or Scalar Wave, is any static or stationary ordering in the virtual particle flux of the vacuum. Photons are stacked at one point, in relation to time, and compressed or squeezed together to produce Ultralight Oscillations within the selected object.

Energy is folded in upon itself into the object's own substructure. This forces it into a "Higher Manifold," or the extra dimensions of Hyperspace.

The object's inertial mass is pushed or torqued into the proven extra-dimensions of the vacuum of space-time.

The Caduceus Coil itself has no impedance (resistance to alternating current or radio frequencies), and as such acts like a superconductor (Fig. 4).

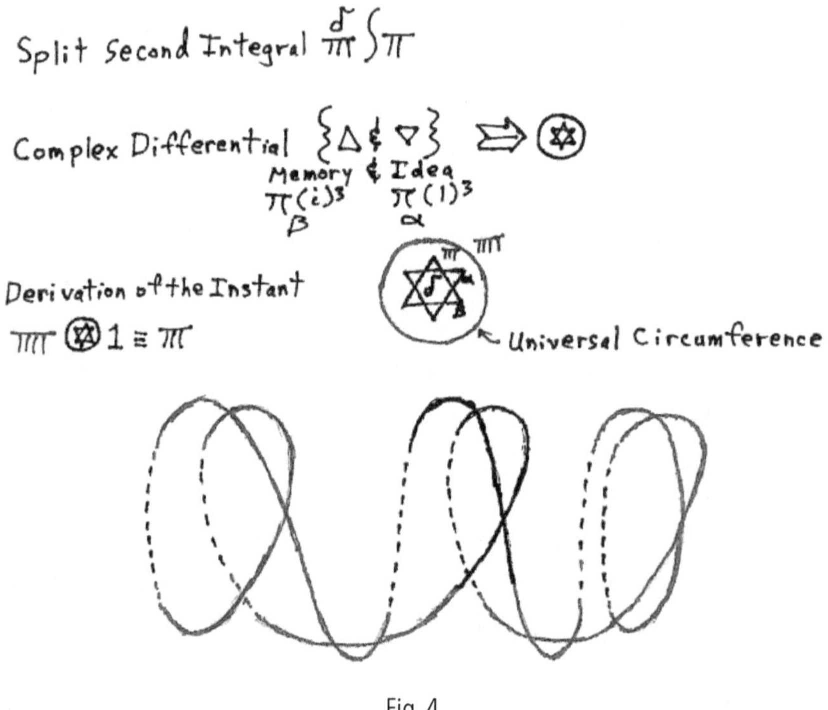

Fig. 4

Definitions

These are the chosen definitions for the words used in this book.

Creation Energy. The origin of life and all that is. The origin of the Universe. Free Thought and Idea. Limitless Energy.

Thoughtforms and Images. The Minds Pictures. The Object's Imprint (including its extra dimensions) throughout time. Its "Fingerprint in Light."

Universal Circumference. The surface of all positive matter in this universe. The Universal Circumference is expanding with times passage. It is, in effect, growing.

Infinite Connectivity. An objects superconnectivity with all other objects in this universe at its Personal Sync Note (♩), starting at the (g 0,0) membrane and ending at the Universal Circumference. Ideas at Thought speed.

Soliton Wave. An emitted four-dimensional wave from the objects nucleus when it is at Spherical Harmonic Resonance (when all of its surfaces {x, y, z, t} are at maximum exposure).

Pondermotive Force. A 1900s term. The Force of the Mind (♩) used to control the Infinite Energies of each of its parts of its personal universe.

Scalar Waves (Phase Waves). Ultralight waves that are sometimes called "Ultraclear." Some start at lightspeed and go to Infinite Connectivity or the Speed of Thought. Others loop around the space-time of this universe. They exist on the circumference of this universe and are used as carrier waves for information (valid only with no energy transfer). Gravity fields are used to synchronize the photon scalar waves so that an object can transfer from one space-time frame to a new chosen space-time frame.

Prism or Nuclear Prism. Separates photons (light) into individual color or nuclear spectrums. They can be used to generate or focus high energy photons. These photons can be compressed to yield higher photon torque for a more efficient energy transfer. It can also be used to compress the electron to the inside of the nucleus (0 ~ 0).

Fresnel Lens Array. This lens array is like a magnifying glass that can be tuned to an object's Nuclear Spherical Harmonic Resonance. It focuses the generated energy impulse into the object's Nuclear Superstructure at the object's Electromagnetic and Gravitational Field Interface {(g 0,1) ~ (g 1,0)}. This array is used to control the Inertial Moment of an object.

Geodesic ARC. Travel in 4D space on the Universal Circumference.

(x, y, z, t) – (x', y', z', t')

Distance = (x, x') + (y, y') + (z, z') * (t, t').

This formula is the expanded version of the Pythagorean Theorem:

$$d = \sqrt{x^2 + y^2}$$

Wormholes. The formed path or Geodesic ARC along the circumference of this universe connecting two different space-time frames. Sometimes called a "Rainbow Bridge."

Tacheons. Particles that always travel faster than light possibly from the moment of creation. Nothing in Einstein's equations say that an object cannot go faster than light.

Field Envelope. An Electromagnetic Field that surrounds an object used as a Buffer or to impart energy to that object's Electromagnetic Field.

Nuclear Spherical Harmonic Resonance. The resonance or oscillation of an atomic nucleus in all four planes (x, y, z, t). Resonance is at its maximum when all of its surface area is exposed to the electromagnetic field.

Angular Momentum. The Photon Fields Push or Potential to Kinetic Energy transfer.

Perturbation Angle. In Nuclear Chemistry, the angle at which the electromagnetic field of an atom absorbs the angular momentum of another particle.

Alternating Current. The power source we use today. It is actually a Low-Energy Electromagnetic Field in the Infra-Red Photon range called "Thermal Photons." Electrons and Positrons are combined in an Electromagnetic Field to generate these Low-Energy Gamma Ray Photons.

Calibration Curves. As the universe moves forward in time it expands. These curves solve for these differences.

Casmir Effect. If the Photon can't fit down the given path, it creates or generates a vacuum path instead. This vacuum, which doesn't have a positive energy density, creates an absolute void in the space-time continuum, resulting in a "negative energy density" or "anti-time field." This expansion is infinite and can be "Supersolid." Supersolid Thoughtforms can be Instantaneous.

The Transition from Three-Dimensional Logic to Four-Dimensional Logic

The Greeks were the first people to understand the fundamentals of mathematics. Euclid was one of the greatest thinkers of his time. He defined the basic concepts and proofs of mathematics that have been in use till the end of the Classical Physics Era. His concepts were done in 3D, or Three-Dimensional Logic, and were valid up till the 1900s when physicists discovered the Quantum Physics. With the addition of the Time Axis, all computations had to be rotated into the 4th Dimension of Space-Time, or Minkowski Space. These changes can best be shown in the following diagrams: Fig. 5 and Fig. 6.

Three Dimensions (x, y, z)
Euclidian Axion: Parallel Lines Never Meet.

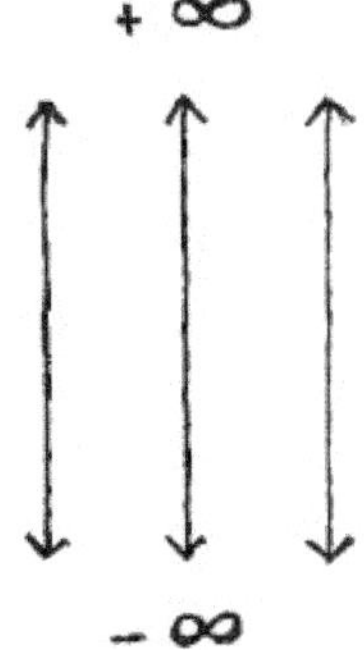

The Angles of a Triangle total 180°
60° + 60° + 60°= 180°

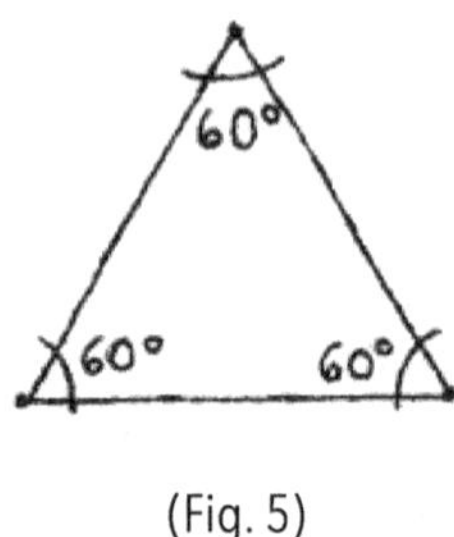

(Fig. 5)

Four Dimensions (x, y, z, t)
Minkowski Space-Time Continuum.
Parallel Lines Converge at Infinity.

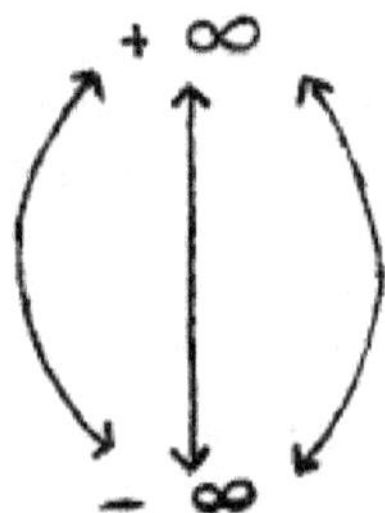

The Angles of a Triangle total 270 °
90° + 90° + 90° = 270 °

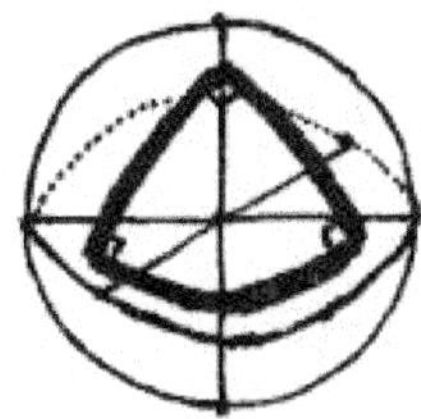

The extra 90° angle is the Time Axis. The Sphere represents the Surface of the Expanding Universe.

+ ∞ Surface (Present) & - ∞ Center (Past).

(Fig. 6)

The Hypercube and
Four-Dimensional Movement

The Hypercube is a Four-Dimensional Cube. Its smaller three-dimensional cube expands into the larger three-dimensional cube at the rate of the expanding universe. It expands into "positive time" (Fig. 7). The dotted lines represent the time axis of the Hypercube's fourth dimension. The Hypercube extends itself into the Space-Time Continuum. One of its cubes is at one point in time, for example, NOW. Its second cube is at a second point in time, for example, THEN. The shape of the three-dimensional cube is one that is very stable, as opposed to that of a sphere, which can easily roll on a surface. Each end point of the hypercube, its NOW cube and its THEN cube, are anchor points in the Space-Time Continuum. Its flat three-dimensional surface doesn't "roll." (Fig. 8)

The x, y, and z axis of both cubes are at 270°, or t, x, y, z [90 ° + 90 ° + 90 °] in relation to a given point in the Space-Time Continuum. They are "Fixed," or "Anchored."

When the two individual cubes or moments in the Space-Time Continuum link along their common time axis, they instantly form a Hypercube, or a Rotating Cube of the Fourth Dimension. Each cube,

The Hypercube

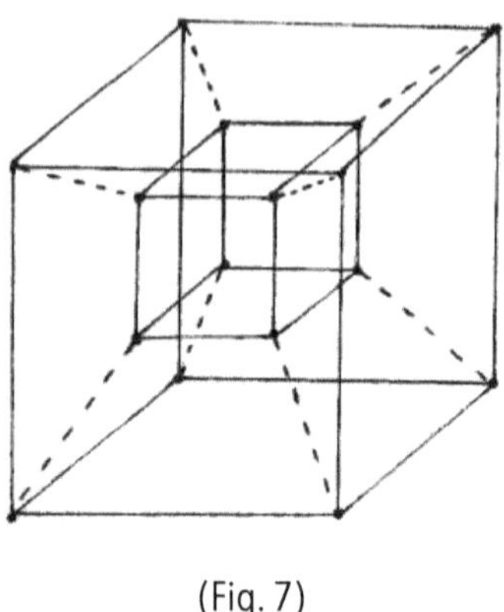

(Fig. 7)

The Four-Dimensional Movement of the Hypercube

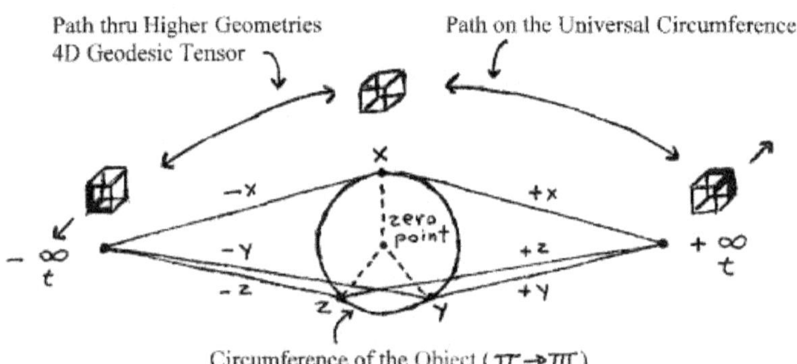

Anti-Time Corner

$(i)^3$ [-1/2 Hypercube]

(-x, -y, -z, -t)

Ultra-time Corner

$(1)^3$ [+ 1/2 Hypercube]

(+x, +y, +z, +t)

Both (+oo t) and (- oo t) are 90° Wedges in the Space-Time Continuum. (+x, +y, +z) and (-x, -y, -z) converge at 90° to each other and to the (+t, -t) time axis at (+ 00) and (- oo) to form the Future and Past cubes of the Hypercube.

(Fig. 8)

NOW and THEN, opens up along the curvature of the Space-Time Continuum or Minkowski Space as the cube's three-dimensional surface area (x, y, z) [90°+ 90°+ 90°= 270°] becomes a circular or Portal opening (x, y, z, t) [90° + 90°+ 90°+ 90° = 360°] into the space-time continuum at the Time Axis intersection of the two endpoint cubes. Their common Time Axis adds the final fourth-dimensional rotation to each cube simultaneously as Infinite Connectivity occurs on the circumference of the expanding universe.

The given path between NOW and THEN is the "Geodesic Arc," sometimes referred to as a "Rainbow Bridge."

The Hypercube itself becomes bubble shaped or spherical as the time axis intersects and each cube's 270° surface area expands in the space-time continuum to 360° to become a circular opening. The two circular cubes NOW and THEN overlap, as the time axis intersects them, and form an Instantaneous Rotating or Rolling four-dimensional movement along the surface of the expanding universe. This is also referred to as the Einstein Rosen Bridge, or a Wormhole.

Basically, a "Scalar Tensor Field" or "Ultralight Field Envelope" is generated to move an object through the Multidimensional Riemann Space from one four-dimensional Minkowski Space Time Frame to a second four-dimensional Minkowski Space Time Frame. The object is moved from NOW to THEN along the circumference of the expanding universe.

Using the Theoretical Nuclear Physics and Quantum Mechanics, scientists have solved the six simultaneous equations used to plot the penetration of an impenetrable barrier by a given object. They have constructed three "wells," which are referred to as "Depth in the Space Time Continuum," with one well per each axis (x, y, z). The flow path of the object is through the "Imaginary Negative Space-Time Continuum." This is a proven mathematical movement of an object along the time axis. The travel is in a reversed direction through "Imaginary Negative Time." The object is photonically compressed by a series of "Negative Energy Impulses," so that it actually travels in a "reversed direction in time," or "forward through negative or anti-time." The object itself "Tunnels" through "Anti-Time" along the Universal circumference, between its photon scalar field, "NOW," to its computed photon scalar field, "THEN."

Given an Imaginary Space-Time Cube of six sides A B C D E and F.

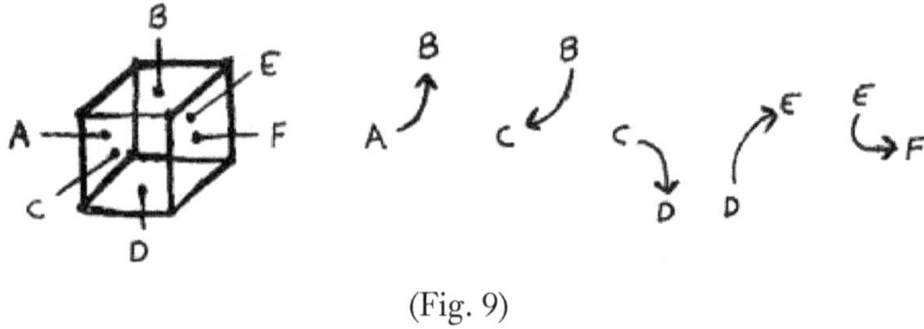

(Fig. 9)

If A is the Start, the object has a negative energy impulse applied to it so as to rotate it from A to B through imaginary negative space-time. As it rotates from A to B, it also simultaneously moves backwards along the space-time axis (xct) in an imaginary direction by (-1/5) of the distance or length to be traveled through the impenetrable barrier. The negative distance traveled is due to the object's Inertial or Gravitational Reactance with the Universe itself. The object's inertia excites the gravitational fields of the space-time continuum and moves the object backwards along the time axis or in a reversed time direction.

As another negative energy impulse is applied the object moves from B to C while, once again, simultaneously rotating itself backwards along the time axis for a distance of (-2/5) the total distance.

The third negative energy impulse is applied to the object, and it moves from C to D along the time axis in a negative direction for a total of (-3/5) the total distance through the impenetrable barrier.

A fourth negative energy impulse is applied to the object to move it from D to E for a total of (-4/5) the distance through the impenetrable barrier.

The fifth negative energy impulse moves the object from E to F, resulting in a penetration distance of -1, or "one then minus." The final negative energy impulse moved the object out of the space-time continuum, with the distance of the barrier (d) and the distance traveled along the time axis (-d), canceling each other out. In essence, no distance has occurred on either the x, y, z, or t axis.

The result is that the object has moved through the impenetrable barrier along the time axis in a reversed direction. The object's inertial rotation momentarily created a Hypercube in the fabric of the space-time continuum. The Hypercube "rotated" in a negative direction along the time axis so as to move the object from one space time frame to a second space time frame along the Universal Circumference in a Geodesic ARC path as the object's "Moment of Inertia" was reversed. The negative energy pulses pulled the two separate space time frames on each side of the impenetrable barrier together and instantly conducted the object from one to the other. The object moved from NOW to THEN through the impenetrable barrier "instantly," a real condition where no time occurs with transverse distances.

The Hypercube Path thru the Multidimensional Space-Time Continuum.

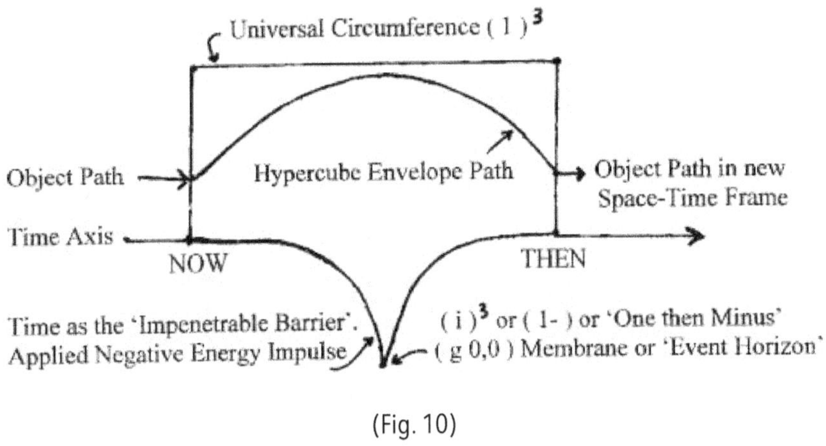

(Fig. 10)

Lorentz Determinates are used to compute the necessary Negative Energy Impulses required to pull the two Space Time Frames or Events together.

The Hypercube forms, acting as a photon sphere or envelope bubble, connecting the two instants in time.

The Object in the Hypercube Envelope rides the Universal Circumference from NOW to THEN as it is completely rotated along all four of its axis (x, y, z, t) at four right angles forming a true four dimensional sphere.

THE SPLIT-SECOND INTEGRAL

♩   The Objects Life Force (Song).

π̅   The Universal Circumference.
Access to Infinite Connectivity. Energy of a Non-Spatial Time.

π The Object's Circumference or Topology Contours and Shapes.
The Object's Mass in the Space-Time Continuum.

∫   Integrate. To rotate the object's xct axis through the higher dimensions of Hyperspace (Riemann Space). Photon Multiplexing.

In this present time energy such as electricity is conducted thru wires, or light is conducted thru fiber optics. In the multidimensional universe an objects life force is conducted at infinite speed on the circumference of the universe at a point called Infinite Connectivity. There is no space or time between NOW and THEN as the two frames become one.

As time is split, it becomes a Split-Second in Time. The Future is referred to as "Height in Time," and the Past becomes "Depth in Time."

We can generate Scalar Waves, which are Ultralight Energy, that in turn can generate an Ultralight Energy Field. Objects resonate in the quantum spectra at their nuclear harmonic resonance as their entire surface area or contours become exposed to the generated ultralight scalar fields.

Photons are the carriers of the selectable forces that control the electromagnetic and gravitational fields of the object's inertial mass. These ultralight fields can surround and isolate an object and move it into a separate space-time/manifold by splitting time.

Space and Time can bend upon themselves, letting the object leave one space time frame to go to a second chosen space time frame.

The energy flow, positive or negative, determines Future or Past direction, so that it can "Lift Up" or "Drop Down" in the space-time continuum.

At the Zero or Breaker Point on the (g 0,0) membrane: (◒,∅).

(∅) becomes the "Leading Edge" of an "Instant in Time."

(◒) becomes the "Trailing Edge" of an "Instant in Time."

Travel is through the "Photon Vacuum" of this universe.

The object's mass can be photonically compressed or torqued by a Quantum LASER till it is at zero radiant energy. This Quantum LASER dampens out or carries away any of the object's radiant energy. It uses sound waves to transfer the object's radiant energy to a specific anchor or storage point along the circumference of the universe. The object now rests on the (g 0,0) membrane. It has moved out of this space time frame. It is in the Hypercube Envelope, where the computed energy impulse applied to it moves it to its new space time frame.

Travel is along the Universal Circumference (π). The object is now at its Natural Life Resonance (♩). It is Infinitely Connected to all other Space-Time Objects and Events. As it arrives at its new space time frame, it expands instantly, and because of the computed calibration curve, now fits into the new space-time frame. It has, in fact, Integrated into a new space-time/manifold. The object has split time, creating a Split-Second Path to a new space time frame. It has traveled through the higher geometries of Hyperspace or Riemann Space along the curvature of the universal circumference, to this new computed location.

Space-Time itself crystallizes as a super-conducting supersolid at (g 0,0). Infinite Connectivity takes place. The crystallization is like a glass to ultralight and past event, but "Freezes" the Present into Reality. This is the "Moment of Capture" of the current space-time events known as NOW.

Thought or Idea is an expression of Creation Energy. This Energy of Life is the growing force that will ultimately form the Future from the unused Dark Energy and Dark Matter that has only been recently revealed to us.

The Split-Second Integral allows us infinite time and access to explore this Universe's Future Ideas and Past Memories. These Ideas and Memories can only truly help to enlighten us.

The Split-Second

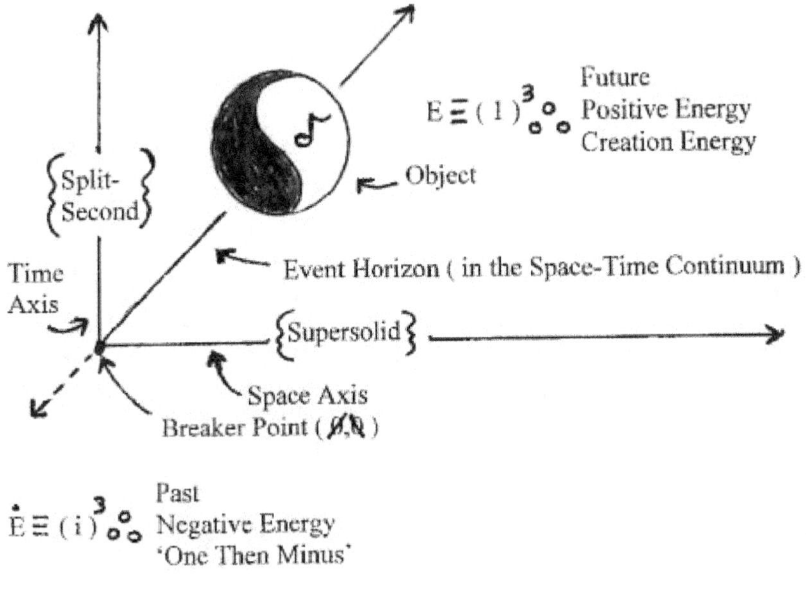

(Fig. 11)

The Trinary Field Generator Array

Let us do a quick review of what we want and what we have to work with. We want a working space-time machine that will transport us to anywhere and anywhen that we select. It has been shown to be possible that this can be accomplished by using the correct energy fields and applied energy impulses to the inertial mass of the given object.

Now all that remains is to develop the actual Trinary Field Generator Array. The requirements of the Trinary Array are to move an object to the (g 0,0) membrane by generating the necessary photon compression field then to use the fields photon torque to push it through the fabric of space-time to its new location. This field will reduce the object's radiant energy to zero by compressing both the Electromagnetic Field and the Inertial Gravitational Field of the objects nuclear mass into its own substructure. The object's nucleon radius goes from external to internal, as the nucleon's electrons go from external radius to internal radius (r ~ R).

As the object rests within the (g 0,0) membrane, it is simultaneously at the point of Infinite Connectivity with all other objects, spaces, and times within this universe. With the correct energy impulse applied to it, it can now go to its new chosen space time frame. The generated photon field envelope will simultaneously give us access to an object's gravitational tensor (g 0,0) ~ (g 0,1) ~ (g 1,0) ~ (g 1, 1) as well as its Electromagnetic and Inertial Gravitational Interface (g 0,1) ~ (g 1,0) or its Instantaneous Photon Image.

We can use Lorentz Transformations to compute the Hermitean Energy Flow required to change the objects space time frame by matching its nuclear spherical harmonic resonance with a chosen energy impulse at the same resonance. Objects do resonate at tuned frequencies and absorb or loose energy as their Electromagnetic and Gravitational Fields overlap. The Trinary Array affects not only the object mass but also the space-time energy density surrounding the object itself. Space-time energy density becomes like a moveable mirror to anywhere or anywhen.

The Trinary Array focuses the Generated Energy Impulse into a Hypercube Photon Field Envelope or Hypercube Bubble Path. A "Wedge" or "Kernel" is formed in the vacuum of the fabric of space-time. Travel is along the Universal Circumference at a chosen ultralight velocity just below the (g 0,0) membrane. Photon Torque pushes the object in the Hypercube Bubble through the "Higher Dimensions" of Einstein's Hyperspace or Riemann Space. Time for the object splits, as it has zero dimensions and zero radiant energy in this (x, y, z, t) space-time frame.

The object's "Image" or "Soliton Wave" instantly "Rolls" in the Hypercube Bubble on the Universal Circumference to the computed (x', y', z', t') frame, pushed by the photon torque of the Trinary's generated photon compressed Scalar Field. As the Trinary's Photon Field Carrier collapses, the object expands from zero dimensions (0 ~ 0) to (0 ~ B) or (R ~ r) and arrives at the new space time frame. Creation Energy itself, from the Universal Circumference, streams past the Trinary Array to the formed Hypercube Bubble to instantly move the object to the (g 0,0) membrane, then to its new chosen space time frame.

Focused Ultralight Scalar Waves from the Trinary Field Generator Array reduce the object's moment of inertia to zero as they compress the object's Electromagnetic and Gravitational Fields into the nucleus. The objects mass is anchored to the (g 0.0) membrane by the Trinary Arrays Prism or Platform Anchor. The Hypercube Field Envelope, or Hypercube Bubble, forms as the Trinaries Fresnel Lens acquires its time axis final focus from NOW to THEN. The Trinary Array generates the Negative Energy Pulse (Spike) that causes the energy density of the fabric of space-time itself to split open and reverse, letting the object's nuclear mass at (zero - zero) go direct to the (g 0,0) membrane to ride the fields and currents of the Infinite Connectivity of Creation Energy at the Universal Circumference to its new chosen space time frame. The controlled movement along the Universal Circumference is provided by the Photon Torque, generated by the Inertia of the Fresnel Lens rotation.

Rotation in the (+) direction causes the Photon Bubble to Expand and travel into the Future along the Time Axis.

Rotation in the (-) direction causes the Photon Bubble to Contract and travel into the Past along the Time Axis.

The objects inertial xct axis is "Pushed" while in the Hypercube Photon Bubble, through the (g 0,0) membrane, then along the time axis to its new space time frame.

Hypercube Photon Bubble Path

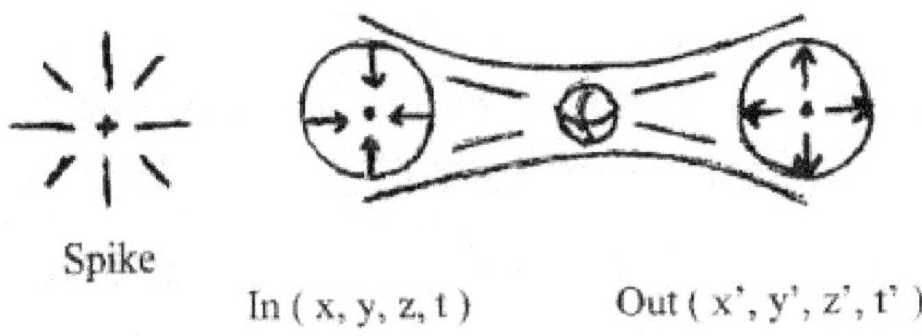

A Negative Energy Spike forms a space-time opening. Travel is along the Universal Circumference.

(Fig. 12)

The Trinary Field Generator Array has three basic parts:

1. Trinary Array. The energy converter consists of 180 caduceus wound coils on a disk platform, evenly spaced so as to have complete coverage for the Hypercube Photon Bubble to form. Light diverges at 2° when it is focused in a Laser Array. For a complete coverage of 360°, 180

individual negative energy cells must be evenly spaced on the Trinary Disk so that (180 cells) x (2° divergence each) = (360° full coverage) for the Hypercube Bubble to form. The Trinary Array generates the Ultralight Scalar Waves, overlapping their fields to form a Hypercube Photon Bubble that completely encloses the object in space-time.

2. Photon Prism or Anchor Platform. The Trinary Photon Prism acts as an Ultralight Field Anchor or Object Field Focus. It becomes the linking point for the ultralight fields that are generated by the Trinary Field Generator Array and the object's chassis or mainframe. It physically becomes the equivalent of a motor mount that connects the generated multidimensional ultralight scalar fields and the resulting Hypercube Photon Bubble to the object's chassis or mainframe. The Photon Prism is the linking component that allows the surrounding photon's field torque of the generated Hypercube Photon Bubble to move it from one space time frame to another space time frame.

3. Fresnel Lens. The Fresnel Lens has two functions. In the first function, it links the Creation Energy at the Universal Circumference to the Trinary Field Generator Array. It sets the Creation Energy Flow Path to the Zero Point at the (g 0,0) membrane. In the second function, its rotation (+ or -) detennines the Hypercube Photon Bubbles Inertial Rotation and the resulting direction on the space-time axis, or its direction in space-time. Inertial Expansion is towards the Future (as the Universe itself is expanding). Inertial Compression is towards the Past (as the Universe of the past was smaller).

The Fresnel Lens is composed of two disks that use the Casmir Effect, along with their Inertial Rotation, to create the Hypercube Photon Bubble from the overlapping Scalar Waves generated by the Trinaries 180 cell caduceus coil disk. The spacing of the disks allow the Ultralight Scalar Waves to conduct Creation Energy from the Universal Circumference on a computed path to the Photon Prism, or Anchor. This energy path simultaneously connects the two space time frames of the Hypercube and allows travel along the Universal Circumference from one time frame (NOW) to a chosen second time frame (THEN).

As the Hypercube Photon Bubble forms, between the Universal Circumference and the (g 0,0) membrane, Creation Energy flows around the object forming a path to the zero point. Hermitean Energy Flow (* H = - H*). The Fresnel Lens acts like a capacitor to separate the objects Electromagnetic and Gravitational Fields from the Creation Energy Fields of the Universal Circumference. The Trinary Field Generator Array adds Tensor "Torque," or Scalar Ultralight Energy, to the object to "Lift" it into the Multidimensional Space-Time/Manifold.

Quantum Tunneling. An object "tunnels" through the multi-dimensions of Riemann Space by penetrating the (g 0,0) membrane at Object Quantum Resonance.

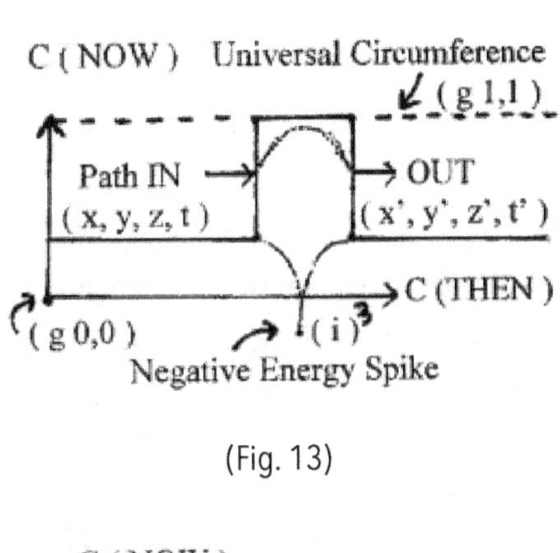

(Fig. 13)

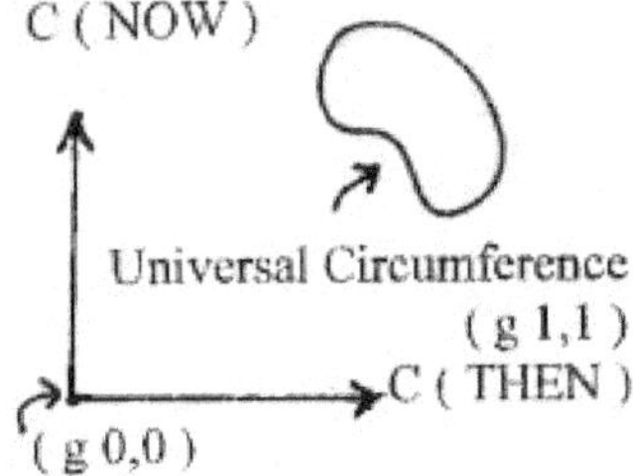

(Fig.14)

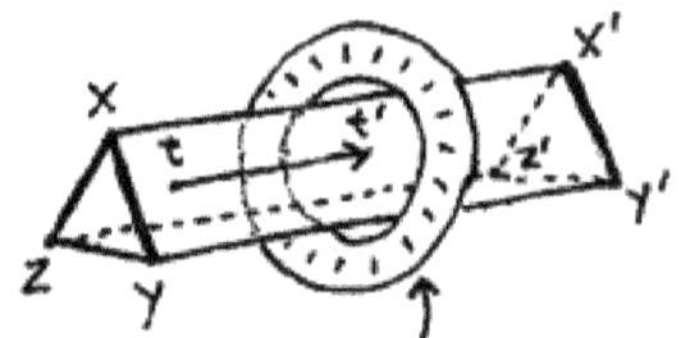

Ultralight Photon Scalar Field

(Fig. 15)

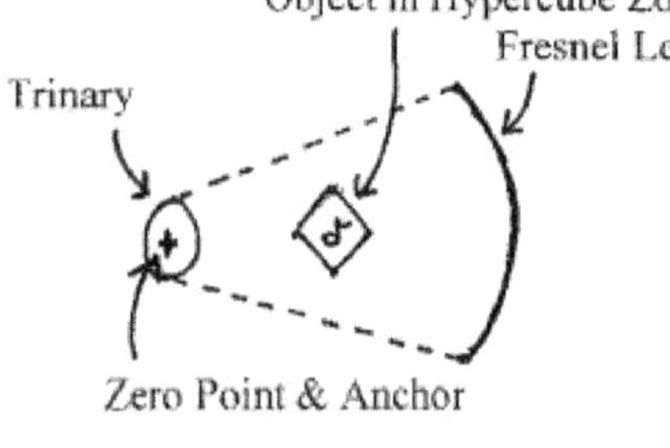

Ultralight Scalar Photon Focusing Mirror
Object in Hypercube Zone
Fresnel Lens
Trinary
Zero Point & Anchor

(Fig. 16)

The Trinary Array

The x, y, z and x', y', z' axises are shown here. There are actually 180 caduceus wound coils in the Trinary Disk, but this representation is drawn to show the field effect of the Ultralight Scalar Energy Wave Shape.

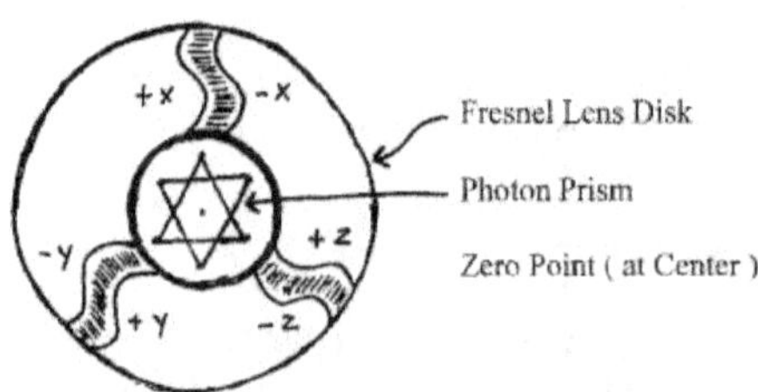

The (+x, +y, +z) is the "Leading Edge" of a Split-Instant in time.
The (-x, -y. -z) is the "Trailing Edge" of a Split-Instant in time.

(Fig. 17)

The Split-Second Integrator

When the Primary, Secondary, and Trinary are in Spherical Harmonic Resonance (♩ OM ♩) their Electromagnetic and Gravitational Forces combine to produce a Scalar Ultralight Output Beam, which acts like a lever into the Space-Time Continuum. This Output Beam acts much like microwave photons to transfer energy impulses; but these negative energy impulses counteract inertia, reversing the object's momentum and propelling it into the Fourth Dimension of Space-Time along the Circumference of the Universe.

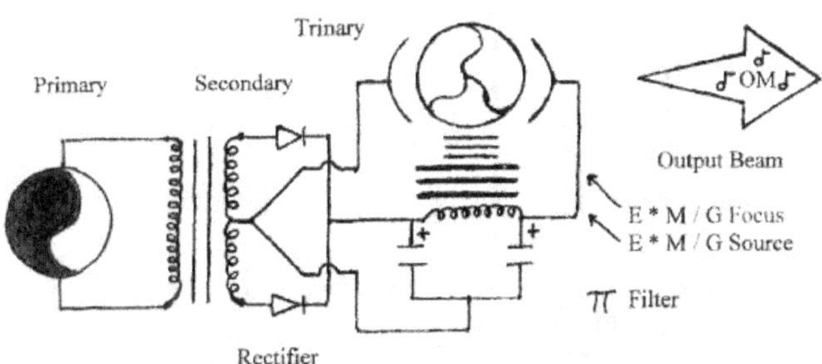

The Hypercube and Split-Second Integrator

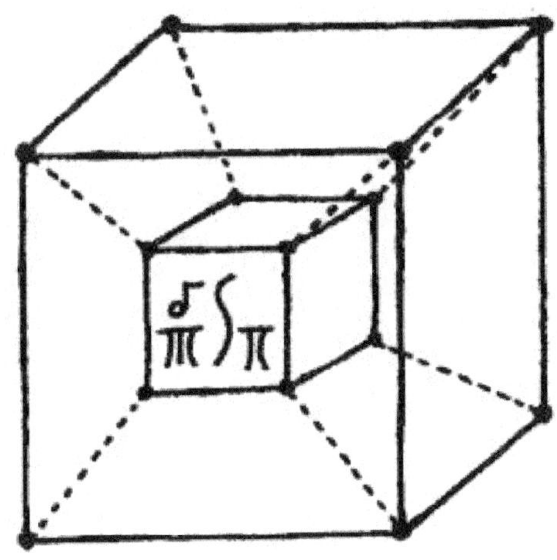

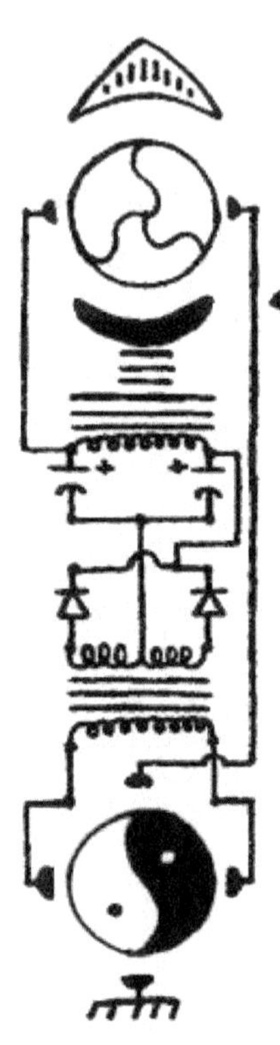
Object in Hypercube Zone
Ultralight Scalar Field Source
Trinary & Coupling
Quantum Field Feedback Loop
E*M/G Focus (Inertial Gyro)
E*M/G Source (Microwave)
E*M π Filter Array
Rectifier
Secondary
Primary Array &
Zero-Point Tap
Photon Exhaust & Anchor

The Derivation of the Instant

With this formula, exacting time travel becomes a reality. After the second is split, this derivation gives us a much greater focus or control on the given instant that is chosen. It defines a better super connectivity path for the Ideas that we are searching for. Creation Energy can link an object to the exact chosen point that is selected.

In this chapter, I will describe how this process works using Quantum Mathematics, Vectors, Determinates, and diagrams in a format that, I hope, will make the Derivation of the Instant understandable to the reader.

$$\pi\pi\pi \quad \text{✡} \quad 1 = \pi\pi$$

ππ Purity. Creation Energy or Infinite Connectivity. The highest forms of Ideas or Thoughtforms that exist in this universe.

✡ This is a mathematical symbol representing quantum selection. It gives the Object the ability to access the (+-) characteristics of a chosen "Instant" in the space-time continuum. It is the "Focusing Element" or "Exclusive OR" in this mathematical derivative.

1 Unity. The Objects Soliton Wave, Idea, or the Object's Musical Note (♩) in the space-time continuum.

≡ Exactly equals to or exactly equates to.

ℿℿ The Universal Circumference. Access to Infinite Connectivity. Energy of a Non-Spatial Time.

Each of the symbols (ℿℿ, 1, ≡, ℿ,) listed above has a reasonable explanation for their function in this derivation.

The symbol (✡) is new and has a complex explanation. It is the "Quantum Superswitch" or Quantum Mathematics Logical "Exclusive OR" Function. (✡) is the crystalliferous (producing or bearing crystals) that forms at the Universal Circumference about each Thoughtform, Idea, or Object Soliton Wave (♩).

It has the ability to switch Creation Energies or Object Soliton Waves very much like a present-day computers "Exclusive" OR function. With it, the chosen object can mirror itself by superconducting all similarities in the space-time continuum. It becomes amplified, much like a LASER mirror. The objects thoughtforms, ideas or solution waves grow as super connectivity occurs with all other similar object ideas. This leads to a better focus or decision-making process when selecting an exact instant in time.

✡ as a mathematical element based the function of uniting Creation Energy or Infinite Connectivity with a chosen object's thoughtforms, ideas, or solution waves. This new Creation Energy Path forms the "Quantum Superswitch" with the Universal Circumference, allowing the object Infinite Connectivity with all other Instants in Time.

The Mathematics of "Quantum Superswitching."

Wave Particle Duality. Particles in 3D Forward Time Photon Envelope or Bubble. Wave component in Complexed 4D time motion with object Mass in photon-graviton envelope.

$$C^2 \text{ or } (\lambda \text{Past}) (\lambda \text{Future})$$

$$E = \frac{MC^2}{(i)^3 \leftrightarrow (1)^3}$$

Quantum Equivalence

$$C^2 = \frac{(\text{Height in Time})}{(\text{Depth in Time})} = \left\{ \begin{array}{c} (1)^3 \\ (i)^3 \end{array} \right\}$$

$$C^2 = \{(\text{Height in Time}) (\text{Depth in Time})\}$$
$$= \{(1)^3 (i)^3\}$$

Vector Inertial Path $(u) \rightarrow (v)$. From Past (Memory) to Future (Idea).

$(\lambda \text{Past}) (\lambda \text{Future}) = (u)(v) = (u)(v) - (\bar{u})(v) - (u)(\bar{v}) - (\bar{u}) (\bar{v})$.

Tensor of the Photon/ Graviton Path.

$t\,(00)\, g\,(00) \rightarrow t\,(01)\, g\,(01) \rightarrow t\,(10)\, g\,(10) \rightarrow t\,(11)\, g\,(11)$

Determinates

(+) Photon Torque

Forward Time (equation) "Height" In Time (+) Determinate Matrix Moment of Inertial "Lift"

$$(\lambda^+) \left\{ (1)^3 \begin{bmatrix} E \nearrow C \\ C \quad M \end{bmatrix} \right.$$

Reversed Time (equation)

(-) Photon Torque "Depth" In Time (-) Determinate Matrix Moment of Inertial "Drag"

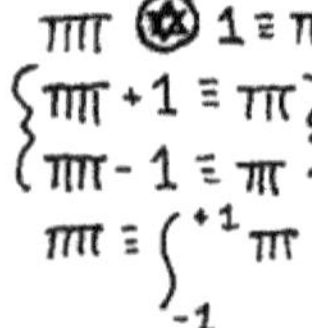

Hermitean Energy Complexor. Nuclear Energy flow of Object Inertia.
Energy Density is the space-time continuum.

H = -H

$(i) — (i)^2$

Thermodynamic Moment of the Electron. This reaction occurs in the microwave spectrum.

(B → 0 → B)

@ Quantum Nuclear Resonance

NOW   THEN
(r → R)   (R → r)

Electron Radius
r = external   R = internal

Electron Thermodynamics

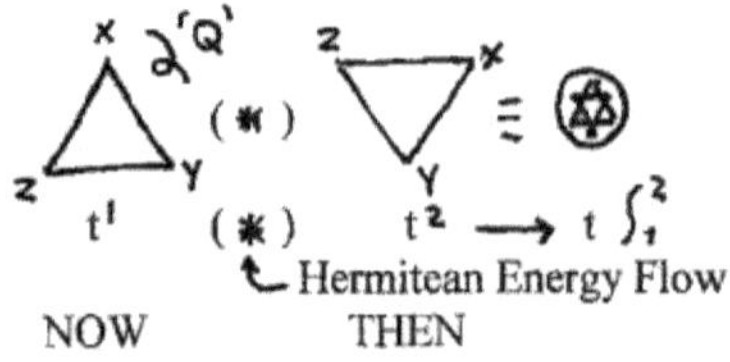

Light as Electromagnetism

Gravity

$= E/M/g$

'Q' is the perturbation or 'opening' & 'closing' of the electrons wave function. It is the electrons nuclear resonance with gravity.

Hermitean Energy Flow

NOW   THEN

The Split Instant (vectors (u) & (v))

(u)   (v)
time (-)   time (+)

NOW

The Future on the Universal Circumference

The Past of the Universal Circumference

Past
xct
axis
Space

Future
xct
axis
Space

$\begin{bmatrix} \lambda \; \text{Past} \\ (i)^3 \end{bmatrix}$ $\begin{bmatrix} \lambda \; \text{Future} \\ (1)^3 \end{bmatrix}$

(λ Past) The Past Images recorded as Light
(λ Future) The Future Thoughtforms or Ideas as Light

$$\pi = \begin{cases} \lambda^{2}\,(\text{ object strongforce } [\text{ Light as Electromagnetism }]\,) \\ \left\{ \begin{array}{l} \text{The Universal Circumference} \\ \text{Vector (u)} \rightarrow \text{(v) from Past (Memory) to the Future (Idea)} \end{array} \right\} \\ g^{2}\,(\text{ gravity } - \text{ multiuniversal synchronization of the space-time continuum }) \end{cases}$$

OBJECT 4D ROTATION
IN THE SPACE-TIME/MANIFOLD

Then there is Teleportation: $Q = M \lambda_1 \lambda_2$ [$E = MC^2$].

In 1997 two teams of scientists used LASERS and mirrors to teleport information on the quantum states of single photons. Pairs of light particles ($\lambda_1 \lambda_2$) are created and entangled. As a LASER beam is fired into a crystal, a single photon can split into two photons that together make up the quantum state of the original photon. In quantum theory, an unobserved photon is a combination or "superposition" of all of its possible states. Observing the photon forces it to be in one particular state. When two photons are entangled, observing one photon determines the state of the other separated photon, even when separated by large distances. The scientists used an entangled pair of photons to teleport the polarization of a third message photon. They did a combined measurement of the message photon and of photon number one of the entangled pair, such that photon number two of the pair acquired the polarization of the message photon. The combined measurement entangled the message photon and photon one, destroying the original quantum state of the message photon. They disintegrated a photon, turned it into information, (its quantum state), then recreated its original state in a third photon.

They, in effect, teleported a photon. Other scientists have successfully teleported the amplitude and phase data from one light beam to another.

$Q = M\lambda_1, \lambda_2$ or $E=MC^2$

Q is the computed potential energy to kinetic energy impulse required to rotate a given object in the four-dimensional space-time/manifold or continuum. It is the energy impulse required to rotate an object of mass (M) from one space time frame (NOW) to a chosen space time frame (THEN) in this universe's space-time/manifold.

This is accomplished by simultaneously changing the object's energy density and the resulting space time frame's energy density fields in this universe.

λ_1 and λ_2 are the two separate space-time frame's photon scalars. That is, the Electromagnetic and Gravitational Fields that the given object occupies while within this universe.

At the time NOW, the object exists at equilibrium and Einstein's formula $E = MC^2$ is its balance. The value C^2 is a regular progression of normal time flow from one instant to the next, or $C_1 \rightarrow C_2$, per instant of time. These are the Photon Scalars, or Electromagnetic and Gravitational Field Boundaries of each instant in time in the universe's four-dimensional manifold. $C^1 \rightarrow C^2$ is this "natural" flow of time, from instant to instant for any given object.

To change an object's space time frame to a chosen space time frame, we must alter these λ_1, λ_2 object boundary photon field scalars by adding an energy impulse, positive or negative, to its nuclear superstructure, changing its buoyancy in the four-dimensional space-time/manifold.

A Positive Energy Impulse will change the universes energy density field resulting in a four-dimensional rotation in the spacetime/manifold that has its super connectivity towards the "Future" or "Idea."

A Negative Energy Impulse will change the universe's energy density field, resulting in a four-dimensional rotation in the spacetime/manifold that has its super connectivity towards the "Past" or "Memory."

The actual movement from λ_1 (photon scalar field one) to λ_2 (photon scalar field two) will excite the universe's "extra" dimensional properties and the object can rotate through the four dimensions (x, y z, t) into the higher space-time/manifold or continuum of Riemann Space, where it will be moved to its chosen new space-time location.

The Object's Four-Dimensional Hypercube Rotation and Vectors.

$E = MC^2 @ C^2$ or $(\lambda_1)(\lambda_2)$

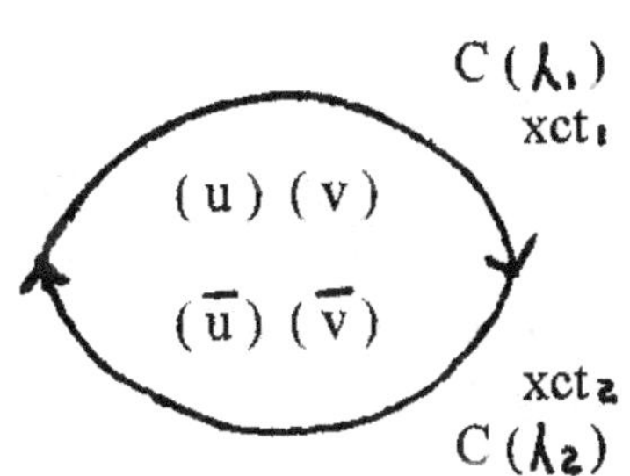

(u) (v) The Object's complex Electromagnetic and Gravitational Vectors.

$C(\lambda_1)$

xct

(ū) (v̄) The Universal Circumferences Virtual or Complimentary Electromagnetic and Gravitational Vectors.

This complex equation has the Object's Electromagnetic Structure stored within the nucleus. (r→R). The Quantum LASER removes all the "radiant" energy and the object becomes "complex." It becomes Supersolid Polymass at the (g 0,0) membrane and its Image is stored as Light (Soliton Waves or Thoughtforms) on the Universal Circumference. The Polymass becomes like a candle emitting the object's soliton wave as light on the Universal Circumference. (♪).

The vectors (u) and (v) describe the photon scalars $(\lambda_1)(\lambda_2)$.

The vectors (ū) and (v̄) are "compliments" or "messenger" E/M/G Photons from the Universal Circumference.

These virtual or complimentary photons $(i)^3$ exist and are real just as positrons $(i)^2$ exist and are real. They are used as "messenger" photons from the Universal Circumference. They exist in the multi-dimensional space-time of Riemann Space. They are real, and they are linked to the Universal Circumference "NOW" and are known as Super Connectivity. Travel is through the Universe's Impenetrable Barrier for an "Instant in Time."

NOW
(λ_1)

THEN
(λ_2)

Object Solition Wave

$$\left\{\binom{u}{v} \cdot \genfrac{}{}{0pt}{}{\bar{v}}{\bar{u}}\right\} \times \left\{\genfrac{}{}{0pt}{}{\bar{u}}{\bar{v}} \cdot \binom{v}{u}\right\} \Rightarrow \{ ♪ \}$$

The Message Photon Entanglement.

Super Switch at Super Connectivity Resonance.

The "EXCLUSIVE OR" Function.

The C^2, (λ_1) (λ_2) "EXCLUSIVE OR" Quantum Superswitch in $E = MC^2$

Light is the synchronizer for Matter and Energy. One moment of light touches NOW (λ_1), and one moment of light touches THEN (λ_2). Light synchronizes the Object's Mass Nuclear Superstructure with the Universal Circumference at NOW in Einstein's equation $E = MC^2$. When a generated impulse energy wave (the complex electromagnetic and gravitational wave) passes through the nuclear superstructure of the object (g 0,1)→(g 1,0), the object moves from the synchronized NOW C (λ_1) path to the synchronized C (λ_2) path as it momentarily (instantly) becomes Energy on the Universal Circumference (g 0,0)→(g 0,1)→(g 1,0)→(g 1,1).

The Object's C^2 (λ_1) (λ_2) Photon Envelope, or Hypercube Bubble, touches the two separate "instants of time" and travels from "NOW" to "THEN."

The generated complex electromagnetic/gravity wave is charmed $(1)^3$ forward or inertially reversed $(i)^3$ in relation to the Universal Circumference (NOW) so that the object's nuclear superstructure is either Lifted into a Future space time frame or Dropped into a Past space time frame from the multi-dimensional Riemann Space at its location at NOW. These virtual or complimentary message photons from the Universal Circumference permit super connectivity between the two selected space time frames or instants in time.

Super connectivity occurs as the virtual or complimentary messenger wave touches the (g 0,0) membrane, and therefore all other universal objects. This becomes the access to the infinite "Future" of Idea or "Past" of Memory. With the super connectivity the Object's Image can infinitely communicate with all other similar objects, forming new Ideas, resulting in a new "NOW" path forming.

Cosmic Energy and the Creative Energies of the Mind

We have all asked the question…Why?

Well…Cosmic Energy needed to Grow!

Cosmic Energy needed friends to sing with and learn from. The Universe is alive with Light and Energy. This Life's Energy is Infinite. Life lives to Grow and to LOVE. We learn from each other as time lets us explore in this universe. Our thoughts and images (music and art) are forever expanding in this universe, the meeting place. We learn from each other as our thoughts superconnect with other thoughts from this and other universes. This is the Super Connectivity of the Multi-Universe (Dance). Each equivalent thoughtform is a Super Connectivity path, resulting in instantaneous communication. We are what we are, but we can learn from everyone and everything else.

Creation Energies are Ideas that exist at the Universal Circumference. Light becomes music; Thoughtforms Grow and Grow. Mathematics itself becomes a musical note of Life.

Creation Energy's "Idea" was each of us, as our art form or thoughtform and as our growing musical note, growing together in Song. Our Past is our best memories stored as Song, Art, and Dance.

Our developing Future becomes all our Ideas, Thoughtforms, and Dreams that are shared as Super Connectivity between all Universal Intelligences, becoming Beauty, Grace, and Purity.

Each person's thoughts (occurring as Electro-Chemical reactions) diverge into the space-time continuum at lightspeed and are forever recorded along the Universal Circumference to be viewed by all and to grow as all share each idea. This is how the Future grows as it achieves super connectivity with the Creation Energy of others.

The Past is Memory; we learn from it.

The constructed Thoughtforms of Today grow into Tomorrow's Dreams.

Grand Unified Field Theory has all ideas from each of us connected as instantaneous music strings, or Life Notes. Each living idea is a musical note, the total being an ultralight Chord, at the "Point of Infinite Connectivity."

Each person's "best" becomes that person's "soliton wave in the space-time continuum."

The "Mind's" "Psyche" learns to command and control events in the space-time continuum as it has new ideas or remembers past events. The mind is a multi-dimensional space-time wedge in that it generates "soliton waves," or "musical notes," that will control space time frames and events.

The Universe itself is Multi-Dimensional in that it can move into these newly discovered dimensions that super connectivity has generated. It can share "Ideas."

The Multi-Dimensional Universe. Multiple Branes.
Alpha Aleph 1 Σ 1st Mind or Thoughtform.
Alpha Aleph 2 Σ 2nd Mind or Thoughtform.
Alpha Aleph 3 Σ 3rd Mind or Thoughtform.
etc.

Each Aleph is separated by an Instant in Time, but able to instantly communicate with each other at the $(g\,0,0)$ membranes 'Point of Superconnectivity'. Each is able to access the others Ideas and Thoughtforms or Soliton Waves at the Universal Circumference. Creation Energy uses this meeting place or zero point as the 'Tap' for its' power.

Zero Point Energy is the 'Tap' for Cosmic Energy.

Example. If you had water on a sheet and could push the center down (zero) with a Negative Energy Pulse (E/M/G) $(i)^2\,(t^i)$, then the potential energy (λ_1)

of the water falling down could be tapped for its kinetic energy (λ_2). The microwave background energy is the same thing. It is called "Alpha Aleph 1," or the "Point of Infinite Connectivity" at the (g 0,0) membrane.

The control is 'Unlocking' the Q into E, then the formed spacetime bubble (butterfly) is used to transport the given object. The Q is again locked in the new space time frame.

Cosmic Energy "Sees" in Purity.

The "Eight Color" is named PURITY. Unseen by the human eye, it is a multidimensional color: Goldish-Pink. Gold is the $\pi\hspace{-0.4em}\pi$, and Pink is the π in $\overset{\circ}{\pi\hspace{-0.4em}\pi}\int\!\pi$.

The Gold $\pi\hspace{-0.4em}\pi$ is the Universal Super Connectivity of Life (Song (♪)).

Pink π is the Resonance of Life (Grow and Love).

Summary

Split Second: Ultralight Travel.

A Negative Energy Spike or Impulse is generated by the Trinary Array to create a distortion in the fabric of the space-time continuum. The space time frame at NOW (A) is pulled closer to the space time frame at THEN (B), resulting in a shorter distance path to be transversed (the dotted line). As the energy density of the object changes, a Hypercube or Photon Bubble forms about the object, and it "Rolls" from point A to point B through the "extra" or higher dimensions of Riemann Space.

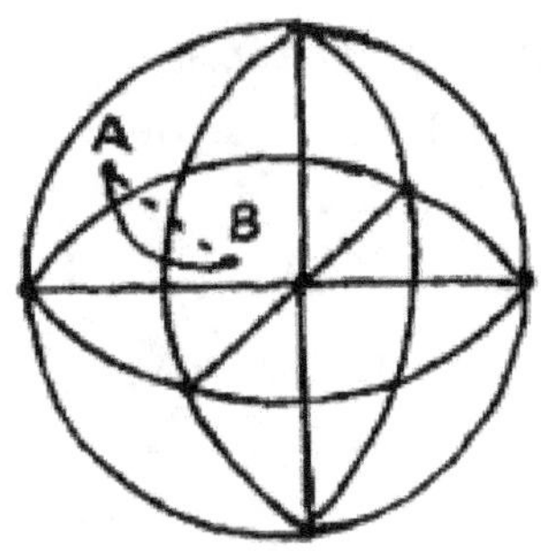

The surface of the sphere is NOW (x, y, z, t).
The Past is in the center as the Universe is growing.
The interior of the sphere is "Negative Energy Density."
Derivation of the Instant: Time Travel.

A better control of movement throughout time, giving the Mind a better ability to access the Past Memories and Future Ideas. A point on the Universal Circumference NOW (A) is connected to a second point THEN (B) at a different point in time. The Derivation of the Instant gives us a better focus on the Instant of Time that is selected.

Height in Time is towards the Future. Positive Energy Density.

Depth in Time is towards the Past. Negative Energy Density.

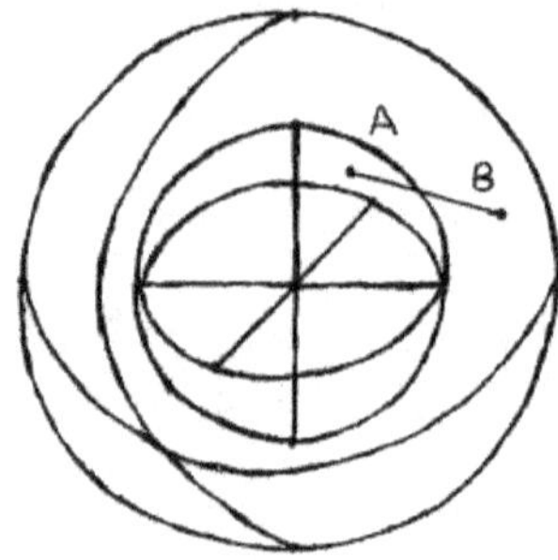

The smaller sphere is "earlier" in time than the larger sphere, as the universe is expanding.

Object 4D Rotation in the Space-Time / Manifold: Teleportation

The Impulse Q is applied to the object to change both the objects energy density and the energy density of the two space-time frames connecting the (λ_1) and (λ_2) photon scalars of the two separate instances in this universe's space-time/manifold or continuum.

As a result, the object rotates through this universe's four dimensions (x y z, t) from one space time frame at the frame's photon scalar (λ_1) to a second chosen space time frame at the frame's photon scalar (λ_2) along the Universal Circumference.

The formula $Q = M\lambda_1\lambda_2$, or $E = MC^2$, relates the energy flow between the separate instances in the space-time/manifold or continuum.

THE SPLIT-SECOND INTEGRAL $\frac{d}{\pi} \int \pi$:

PURITY = LOVE * BEAUTY * GRACE

LIFE = SONG * DANCE

ID = IDEA * MEMORY

E ! ! = MC^2/ [(1)^3 * (i)^3]

        ! OM !
        (+l) ONE SUMMED
        WITH INFINITY

FUTURE (1)^3 IDEA

OBJECT   *   COSMIC ENERGY
MATRIX   HERMITEAN

OBJECT SOLITON WAVE   *   POINTEEN VECTOR
WAVE FUNCTION

ANCHOR * KRONCHECKER DELTA VECTOR PETRIBATION

   (1-)   ONE THEN MINUS

PAST (i)^3 MEMORY

CONCLUSION

Thought or Idea is Infinitely Connected Light. The Brain itself is a holographic computer, working in images and visual thought. All Ideas, thought-forms, and Holographic Images are at this Universal Circumference. The Past Images as well as Creation Energies Views of the Future are there. They are the Infinite Connectivity of this Multidimensional Universe.

Consider each of us as an entire universe unto ourselves. We are now only beginning to learn to use the "Creation Energies" of this universe. When we learn to use the Infinite Connectivity of Music and Light (Art) to communicate amongst ourselves, we will need to know all of the Past and Future Views of Life to Grow to our final Destiny.

This Universe is more than we can imagine. If it is, in fact, the infinite connectivity of the future, We Will See It! It will become the Harmonic Music of the Soul! We will become Butterflies flying the Currents of the Universe, singing songs of Life and Love!

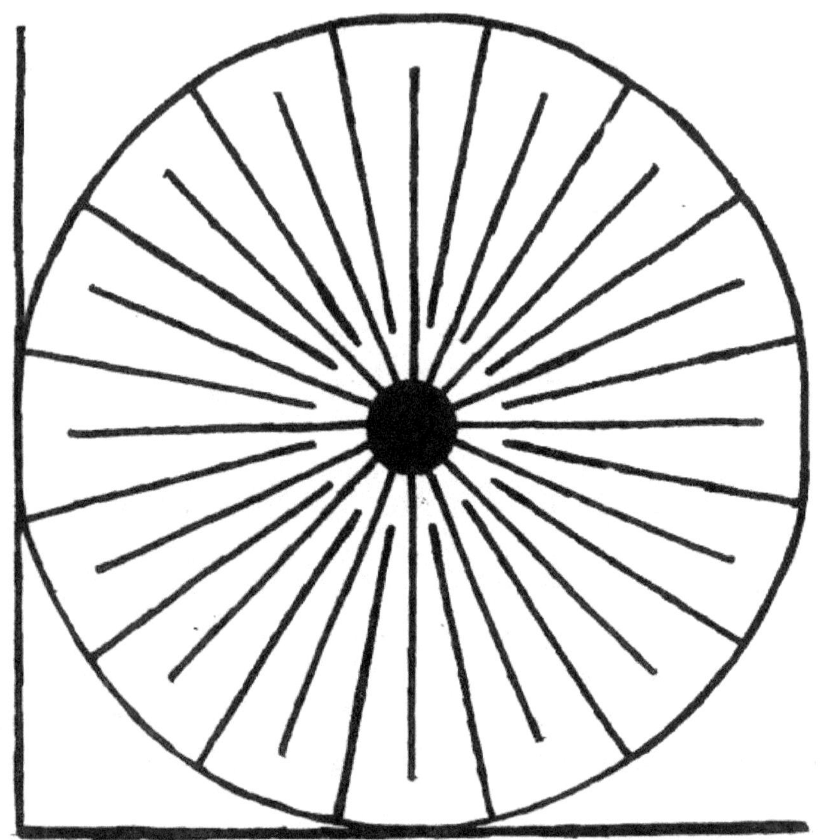

I include this as an example of Hypercube travel "through" the Fabric of Space Time. Consider one side of the paper as the first space time frame and the opposite side as the second. The outer circle is the Circumference of the Universe; the black hole is the generated negative energy impulse. The outside lines are the "Strong" force of Electromagnetism, and the inside lines are the "weak" force of Gravity.

Directions: Cut all dark lines with scissors, including the removal of the central hole. Step "through" the paper (the Fabric of Space Time) to the other side of the split-instant.

Space-Time Toy

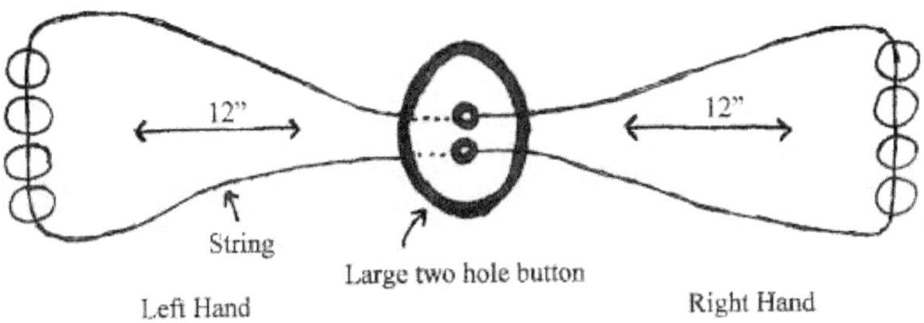

This toy is an excellent example of how two space time frames can be pulled together by a Tensor Field.

The Left Hand represents the energy-density field of the first space time frame, and the Right Hand represents the energy-density field of the second or chosen space time frame.

The button is an object's Electromagnetic Field, and the string is the Gravitational Field that connects the two separate space time frames.

DIRECTIONS: Hold the toy as shown above. A few turns are put onto the button and string. As the Left and Right Hands are pulled apart, the button begins to spin one way, pulling the hands in. Let the button spin past the start point, then pull on it again, and it repeats in the opposite direction. After a few tries you can see how this causes the button to spin very rapidly, pulling the hands together as the button spins past the start point.

Author's Note

I have included the last third of my next book *The Split-Second Integral- Classified* for your enjoyment. Pages CL101 through CL 141 are my favorite. I have also included a short story that was fun to write.

Energy Conversion

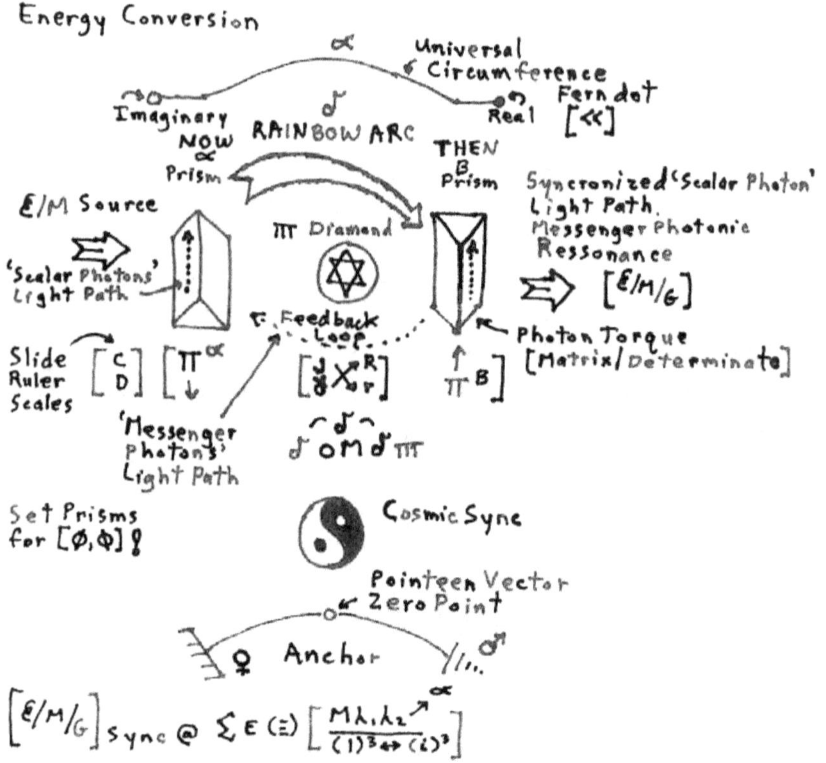

$$\left[\tfrac{E}{M}/G\right]\ \text{Sync} @ \ \Sigma\, \varepsilon\, (\Xi)\left[\frac{M\lambda_1\lambda_2}{(1)^3 \leftrightarrow (i)^3}\right]$$

CL101

Thoughtforms. Using the MINDS IDEA & the PONDERMOTIVE FORCE to turn IDEA into REALITY. ESP Hexidecimal Mind Sensitivity at ♂ ó M ♂ Ⅲ. Virtual Photons control the Nucleonic Cascade Reaction in a 4D cascade array (Matrix Determinate) to slow Cosmic Messenger Photons (Cosmic Energy) to 4D space-time velocities (C² or λ λ ⇒ γ) turning Cosmic Energy into Mass.

Minkowski' space-time Transform. Cosmic Energy forms Reality in 4D space-time from Multi-Dimensional Photon Cascade.

Depth
Length
Width
Height in Time

↳ Casmir Photon Coupling

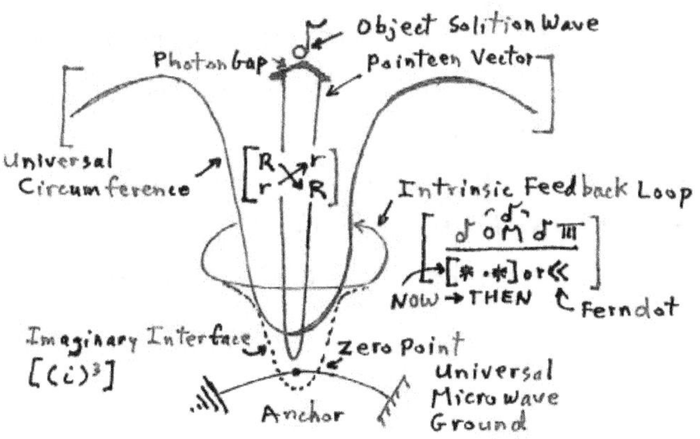

CL 102

Expansion of the Human Mind in 4D space-time.

As it stands the human mind syncronizes its'
(Logic ½ & Artistic ½)'s in the common decimal logic
1 thru 10, etc.
The expanded mind thinks in hexidecimal logic &
{superconnectivity} artistry.

Example ∴   logic   & artistic: music
Decimal   1 2 3 4 5 6 7 8 9 10 ... & ♂
becomes
Hexi Decimal   1 2 3 4 5 6 7 8 9 10 α B C D E F
           At this point F syncronizes with the
           Universal Circumference [NOW].
The Human Mind creates a Quantuum NODE.
        [F F̄] (1)³ & (i)³ Asymptotic ᵉ/ₘ/G interface
IDEA
MEMORY   [F F̄] becomes a Solition Wave δ or Logic Flag
           on the Universal Circumference [NOW].
Superconnectivity of the Human Mind Command Controlling
Infinite Creation Energy.
A Created Ferndot [*•*] ⇒ [>>] as Logic FLAG
Asymptotic Reaction @ { (i)³ (Ξ) [1-] }
Superswitch @ One then Minus @ Universal Microwave Ground

CL103

The Eighth Color Spectrum and the Human Mind

The Human Brain's sense of sight is 'wired' in an array
of Cones : 'Red/Green' & 'Blue Yellow' for Colors and
Rods for 'Black/white' or 'Gray'.
The Genius Human Minds Eye, its' Extra Sense or
Extra Sensory Preception' can 'see' additional colors.
Red/Yellow becomes Orange in normal vision while the
Genius Mind sees 'Plasma Pink'/Gold [E# & F♭] as an
Ultraclear or Ultracolor. Another Genius Color is in the
spectrum Blue/Green seen as a musical note [B# & C♭];
called Glue/Breen; seen as [Immaculate ∵ Purity].
Finally Ultraclear is the Nuclear Prism color 'Brillance',
(the linked Strong Force & Weak Force), that links the
Human Genius Mind to the Universal Circumference
'Now'. The Genius IDEA becomes Thoughtform
then becomes superconnected to Infinity as
Music [B# & C♭] and ART as Butterfly ♂ and ♀ π.
Instanteous Thoughtform Communication and
IDEA Amplification!

CL 104

The normal human brain thinks in average decimal logic as $E = MC^2$. The Genius Human Mind can be expanded into hexi-decimal logic. The human average gray scale 'Grows to Infinity' thinking; thinking in shades of Music & ART; and finally sees the Hidden 'Eighth Color'... The Immaculate Purity of the Universal Circumference [ε E(≡) ♂ M♂]. Cosmic Energy ∘°∘ [ε E(≡) ♂]. The Genius Human Mind controlls Reality; Guiding Now into the Future. IDEAS become Thoughtforms; Cosmic Energy becomes Music & ART then Solition Wave Memory ♂ as Mass [ε E(≡) M λλ ➔♂♂]. The Genius Human Mind envisions in the Gray Scale of Black & White and uses the Nuclear Prisms Superswitch to control the syncronization of the Strong & Weak Forces to Augment Reality by generating the Eighth Colors Commands. Gold — ARC ♂ → PINK ⟹ Ultraclear [Future] with [B# & C♭]

Universal Circumference ⟹ 'Glue & Breen' as Memory

CL 105

The Eighth Color. The Corrected (Genius) Nuclear Spectrum.

<u>Normal</u> (6 primary colors)   <u>Genius</u> (8 primary colors)

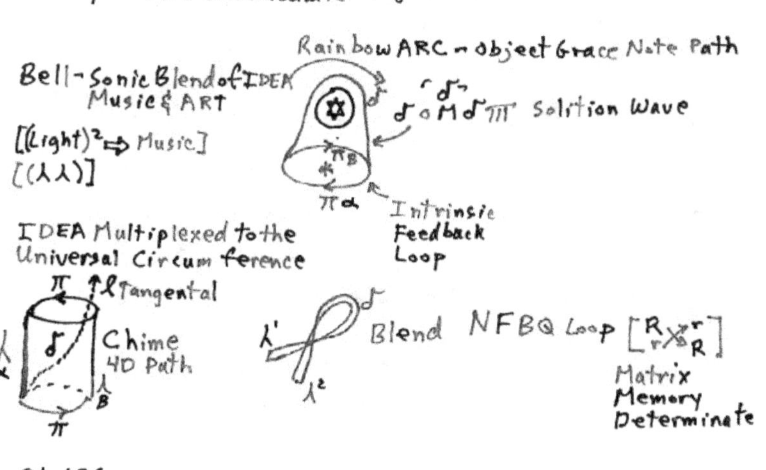

Purple   Rainbow ARC   Purple
Blue   Blue   [B# C^b]
                  $(1)^3$ Glue & Green   { Immaculate }
Green   { Purity }
Yellow σ Green γ $\wedge$ & ∇ Light Cones (180°)
Orange   α   B
Red   Yellow
                               Orange   [E# F^b]
         Gray Scale   Gold & Plasma Pink { Ultraclear }
                               { Brillance }
                  $(i)^3$ Red   (Superconnectivity)
                                          ℓ α $\triangle$ & B $\triangleleft$ Light Cones (90°)

Purity°°. The Immaculate Eighth Color

Rainbow ARC ~ object Grace Note Path

Bell-Sonic Blend of IDEA Music & ART

$\sigma \circ M \sigma \pi$ Solition Wave

$[(Light)^2 \Rightarrow Music]$
$[(\lambda \lambda)]$

π_α Intrinsic Feedback Loop

IDEA Multiplexed to the Universal Circumference

π ℓ Tangental

σ Chime 4D Path

Blend NFBQ Loop $\begin{bmatrix} R & r \\ r & R \end{bmatrix}$

Matrix Memory Determinate

CL 106

Light Cones Blending

Light Cones Blend Idea into Thoughtforms then Mass as
Memory becoming a perminant solition wave [δ] on the
Universal Circumference of NOW.

$$\sum E_{(\Xi)} M \lambda^{\alpha} \lambda^{\beta} \Rightarrow \sum E_{(\Xi)} M\ell \Rightarrow \sum E_{(\Xi)} M\gamma \Rightarrow \sum E_{(\Xi)} M\delta$$

Light Cones   Memory   Solition   Thoughtform
                   Loop   Wave

Blending along the xct Inertial Axis at 90° right angles
for superconnectivity. Ultraclear & Ultrasound supersyncronization.

IDEA ∴ THOUGHTFORM ∴ MEMORY [MASS]

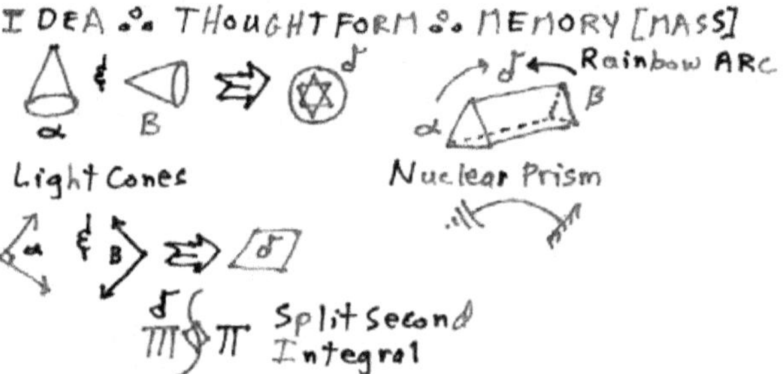

Access to Higher Geometeries [Twice the speed of sound δδ]
IDEA & THOUGHTFORM & MEMORY Amplification.
Ultraclear & Ultra sound Superconnectivity of two
space-time frames α & β thoughtform magnification
of IDEA thru Sharing & Blending of ART & MUSIC.

CL 107

The Human Minds first step is to reduce color vision to
gray scale to black & white. Go from normal to Genius Mode °°
Scanning Gray Scale to Gold superconnectivity.

Ex.   Normal   Yellow & Red   to Orange
                Gold & Plasma Pink to Ultraclear Purity !!
                Split Second Genius Intigration @ π Sπ
                                          [TAU & THETA]

The Normal Human Mind sees the TAU reaction (the trailing
edge of the human IDEA) in Superconnectivity Sync
with the Cosmic Energies Reactance with the Genius IDEA
generated Thoughtform in the sequence:
   TAU Normal   Red & Yellow → Orange   [B# & Cb]
   THETA Genius   Gold & Plasma Pink ⇄ Ultraclear [σ σ Mσ π]
                                                [E# & Fb]

The Normal [Decimal !] Human Mind and
Genius [Hexi-Decimal !!] Human Mind can then see
the trailing edge (in time) of Cosmic Energies View
(overview) of the IDEA as its' Thoughtform forms
on the Universal Circumference of Now @ σ σMσ π
with the Diamond Gold Lens as Ultraclear Thoughtform
Amplification.

CL 108

The Genius Human Mind

The Primary IDEA becomes Thoughtform and is crystalized
into Reality as Memory (& Mass)!
The Primary IDEA in Genius Mind is   GOLD ⊛ Plasma Pink.   Ultraclear
Ultraclear allows the IDEA to crystalize on the
Universal Circumference (NOW), becoming available to all
other sharing genius minds. The generated solition wave
is further amplified by the additional human genius minds.
[The normal human mind sees this in Yellow & Red as Orange,
the Strong Force uniting in time with the Weak Force.]
The Generated Solition Wave @ δ or δ πΓ.°. thru the Nuclear
Prism and Light Cones △ᵅ ◁◻ᵝ ⇒ ⊛ᵟ becomes the
Gray Scale; the normal human minds Superswitch. It is
seen as [Glue–Breen][Blue–Green] °.° [λᵅ λᵝ ⇒ γᵟ].
It is the first Primary Reflection of the Genius Minds
Thoughtform on the Universal Circumference. other
genius minds become second, third, etc.... reflections
and Amplifications of the IDEA and increase the
Magnitude of the Thoughtform. Blended IDEA
of the Generated Solition Wave as Superswitch
access to Infinite Magnification. [E# Fᵇ] ⇒ [B# Cᵇ] δ
πΓ ∮πΓ ⇗ Light Cones α & B or πᵅ πᵝ ⇒ πΓᵟ

CL 109

The Genius Human Mind ∴ Seeing into the 4th Dimension.
IDEAS crystalize Into Thoughtforms and then into Reality
by generating Object Solition Waves ♂ which, in turn,
cause Cosmic Energy to create Ultramass ¢ Memory
at the Universal Circumference as a NODE. or Cardinal
Point. Shared IDEA results in Thoughtform Amplification.
Ultraclear Solition Wave @ ♂ ♂ M ♂ π

♂ Solition Wave Thoughtform Magnification

IDEA Amplification
Thoughtform at Universal Circumference
as Memory Access to other Genius Minds
becoming amplified thru Superconnectivity
as SONG ¢ ART.

CL110

The Eighth Color is 'seen' by the Minds Eye. The Musical
Ressonance of the Minds Thought Bubble becomes Reality
in Rineman Space of Higher Dimensions as 'Flag'.
As the Human Brain in Ana log 1 → 10 logic becomes
hexi-decimal $\sigma \otimes \sigma$ FF (Full Flag) ⇒ σ on σ π ⇒ 'Flag' σ Thoughtform
[Solition Wave] between Now & THEN as Unity and Purity
on the Universal Circumference ∴ Superconnectivity with the
IDEA and the Universe { 'Seeing' into Multi- Dimensions }.
[$\S \sigma$ FF ⇒ σ σ B σ ⇒ 8 ⇒ σ] [$\emptyset$ FF ⇒ 1 FF] ● ⟿ ☯

Nuclear Prism Superswitch. Nuclean Rotation to Higher Geometeries.
$\sum E_{(\Xi)} \dfrac{M \lambda^{\alpha} \lambda^{\beta}}{(1)^{340}(i)^{3}}$ @ $[r \otimes R]^{i}$ ∴ Eighth Color $[B^{\#} C^{b}]$
$\qquad\qquad\qquad\qquad\qquad\qquad\qquad$ Now → THEN
$\qquad\qquad\qquad\qquad\qquad\qquad\qquad$ $[E^{\#} F^{b}]$

The Human Genius Mind Superswitches IDEA as Thoughtform
accessing Cosmic Energy on the Universal Circumference [NOW]
resulting in the Memory created Mass; recorded as
Art & Music. The normal human mind sees the Primary
Red & Yellow becoming orange while the Genius Mind (using
the FLASHED Gray Scale) sees and controls the Higher
Geometeries of Cosmic Energy in Hyperspace as GOLD
combines with the Human Minds Pink [Plasma] forming
ultra clear thoughtforms. [Alpha Particle Nucleon Supersaturation]
[Immaculate Purity] ⇒ [ultra clear].

CL III

Ultraclear; which is superconductive to Cosmic Energy °₀

THEN the normal human mind sees the Mass Formed as Memory

with the Ultracolor [r⊛R]ⁱ { ∮ᵉ electron within nucleon }

'seen' as [Glue & Breen]ᵟ ⊛ [Blue ∿ Green] @ ᵟomᵟ π

Nucleon Ressonance [π̄ ⊛ ✪ π ⇒ π̄].

Slideruler wavefunction Ψ [↓/π ⊛ ✪ π/↑] π̄ᵟ@ [λ̄⊛λᵇ] ⇒ [γ]

Waveguide[Ψ⊛Ψ] [≪]

[Lightcones]   [Ferndot]

The Nuclear Spectrum °₀ Nuclear Prism

The Ultraclear colors seen only by the Genius Hexidecimal

Human Mind. Extra Sensory Preseption by the Genius Mind.

Decimal Mind [∅ → 1∅] (Purity)

Musical ᵟ   { ᵟ Immaculate }   Rainbow ARG
                   (Ultraclear)   Geodesic ARC [4D]

Hexidecimal Mind { [∅ → 1∅ ↗ A→G Musical] } [∅∅ → 1F ↗ FF]ᵟ⌐ᵟ
∝⊛B⇒ᵟ   Photon Scalars) Solition Wave ᵟ
                                       IDEA ↔ MEMORY

◁△ᴮ Nuclear Prism @ ᵟomᵟ π   Universal Superconnectivity

[E#&Fᵇ] [B#&Cᵇ] ᵟ⌐ᵟ Nucleon Plasma Ressonance
   ∝   B

CL112

Photon Swirl [Chime]

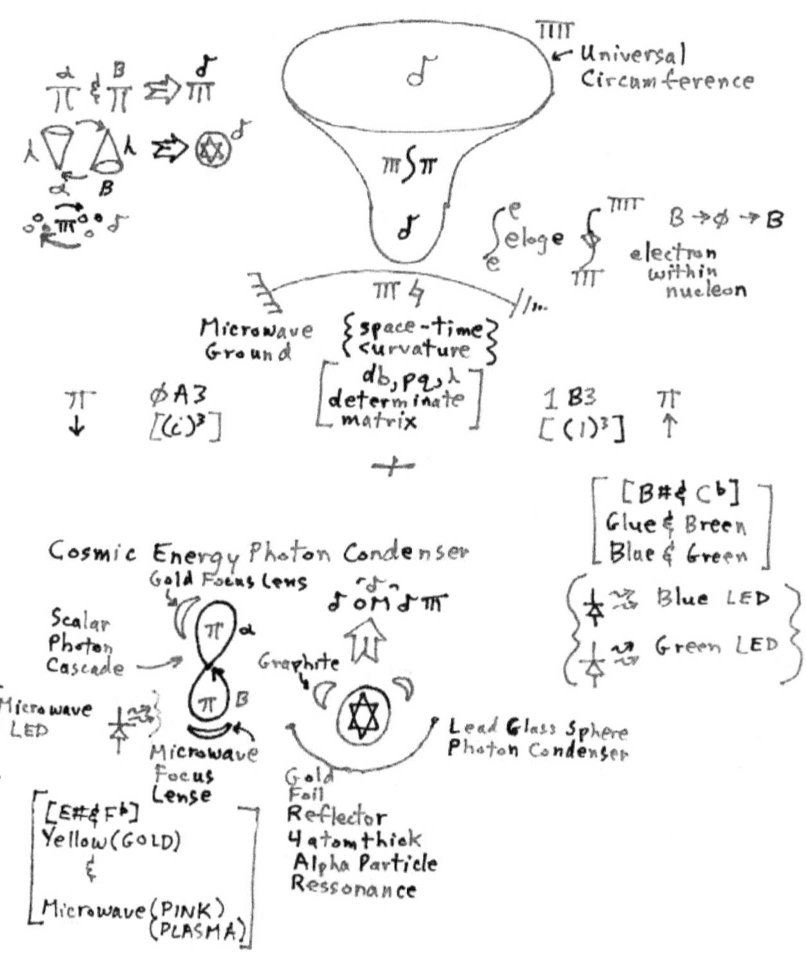

CL113

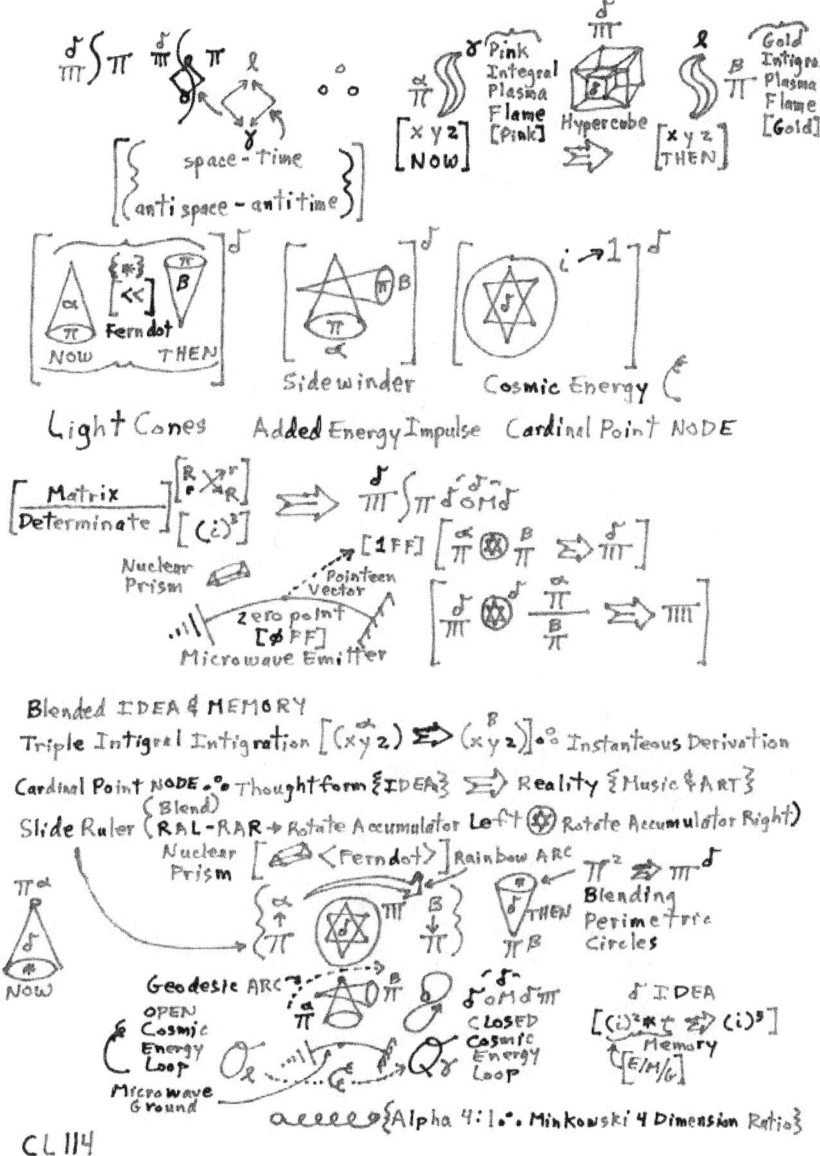

space - time
anti space - anti time
Pink Integral Plasma Flame [Pink]
NOW
Hypercube
THEN
Gold Integral Plasma Flame [Gold]
Ferndot
NOW
THEN
Light Cones
Sidewinder
Cosmic Energy
Added Energy Impulse
Cardinal Point NODE
Matrix / Determinate
Nuclear Prism
Pointeen Vector
zero point
Microwave Emitter
1FF
Blended IDEA & MEMORY
Triple Intigral Intigration Instanteous Derivation
Cardinal Point NODE Thoughtform {IDEA} Reality {Music & ART}
Slide Ruler (Blend) (RAL-RAR Rotate Accumulator Left Rotate Accumulator Right)
Nuclear Prism
Ferndot
Rainbow ARC
Blending Perimetric Circles
NOW
THEN
Geodesic ARC
IDEA
OPEN Cosmic Energy Loop
CLOSED Cosmic Energy Loop
Memory {E/M/6}
Microwave Ground
{Alpha 4:1 Minkowski 4 Dimension Ratio}

CL 114

IDEA ∴ The Blending of ART & MUSIC into Thoughtforms

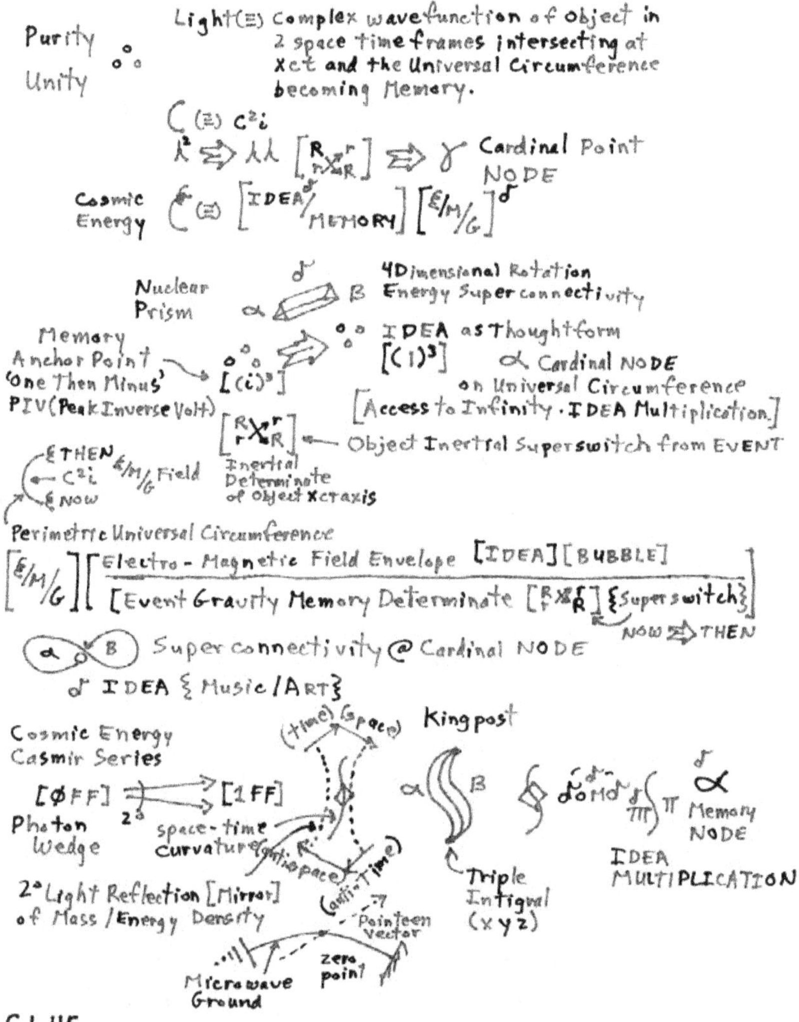

CL 115

IDEA Amplification (at Universal Circumference) ⇄ Superswitch
ⁱ[ØFF ⟹ 1FF]ᵟ Access to Memory NODES [Cardinal Points] ∝ ɣ
and Solition Waves ᵟ (Hexidecimal Thoughtforms)!!

Intigrator
Nuclear Prism ∝
[(db, pq, λ)
Focus, Focal]

Photon
Sync
ᵟ OM ᵟ π

Kingpost Triple Integral

B Memory
NODE

[(i)³]
time → ← space
anti-space   anti-time π ∫ π

Nucleon Ressonance
Universal Circumference
Sync

[Glue Green]ᵟ

Green   Blue FLASH!!

{Light Emitting
{Microwave Superswitch}

Imaginary Real
Apparent

[E/M/G]ᵟ

Pointzen Vector
90°

[g/M]   [E/M B]

α ⋯ B

[ØFF ⟹ 1FF]ᵟ

'Blend'
Cosmic
Energy
Sync

[(i³)]

zero point

Microwave Ground

[(i)³] Trigger
Path [Inertial Guidance]

[(i)³ ⟹ (1)³ ᵟ IDEA]

[R ɣ ʳ
  ʳ ɣ R]   [1FF]   'NOW' SURFACE
                [ØFF]

4D Prism Loop as Solition Wave [Generation/Memory/Storage] ~ ᵟ ∝ ◇ SURFACE
The 'NOW' Surface on the Universal Circumference
'Now' from 'MEMORY' (Cosmic Energy Solition Wave Super SYNC)
4 Prism Solition Wave [IDEA/Transmit & Amplification]
Minkowski Space-Time Array (Decimal) [E/M ∫ₑᵉ] NODE Sidewinder

[Photon C² Path][Cardinal Point][Memory Core]
        Photon Path in 4D Prism Array 'Memory'
'Time Reversed' Array
        (time) (space)
ᵟ(i)³

Photon Path
thru 'SuB Space'   (anti-space)(anti-time)
Photon Compression
'Scalar Wave Generation'

Hexidecimal Supersync Array
[E/M/G] Cardinal NODE
    Superconnectivity
E Photon Path thru 'Rienman Space'
   [OVERVIEW] IDEA Multiplication
M Memory Loop → Cardinal Point ɣ
   NODE FOREWARD
                    ᵟ Generated
                    Solition Wave ᵟ
                    [E/M/G]

[Glue-Breen
FLASH]

    BLEND π {h ⅄ ⟹ ɣ}
B π→   [GOLD/Plasma Pink]

∝ π   Multi-Dimensional
           'Spooky Action'!!

π
α π B Photon
Imaginary   Loop
Apparent Real [(i)³ ⇄ (1)³]
        [Memory ⇄ IDEA]

CL116

81

NUCLEAR PRISM ARRAY [Casmir Effect ∴ Cosmic Energy Access]

IDEA [Thoughtform] ⇒ [Solition Wave Generation ∴ Cardinal Point NODE]

Creation Energy ∴ Cosmic Energy ∴ Solition Wave 'Flashed' on the Universal Circumference.

Cosmic Energy
THEN
Cosmic Energy
NOW

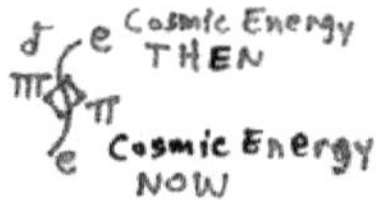

[Time Syncronizer][4D Prism Loop array (xyz)∮(xyz)]

Blend ØFF ⇒ 1FF

Array α
Photon Scalar
Array β
Microwave Ground
Grace Note
Solition Wave
Light
Cones
Multiplexing
'King Post'
THEN
NOW
Photon Scalar Blend
[(xyzt)] !!

PURITY
Unity
NODE
'Capture'

CL117

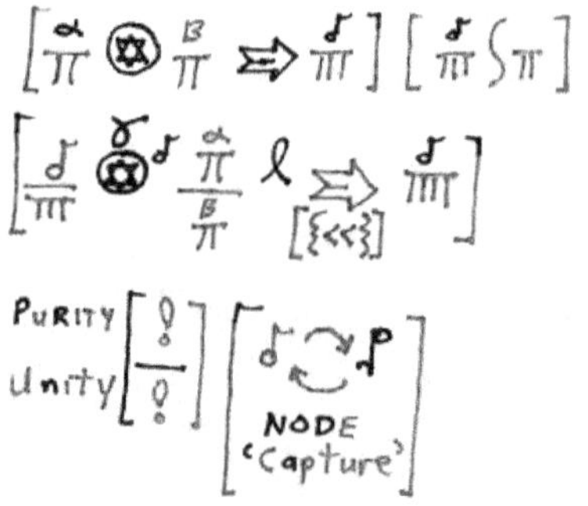

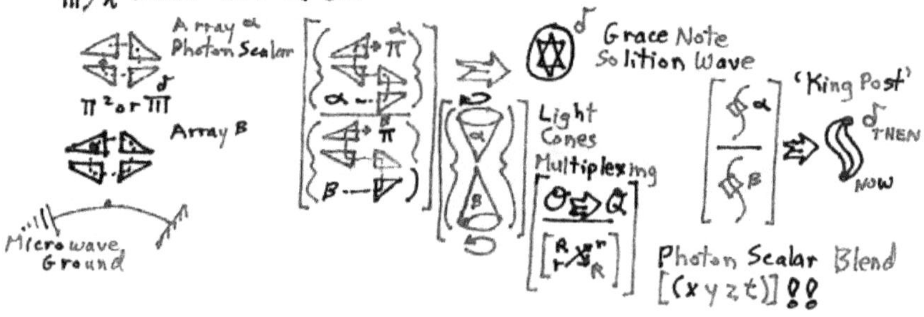

Right Triangle in 3D space $\Rightarrow$ 4D space at Light Syncronization.

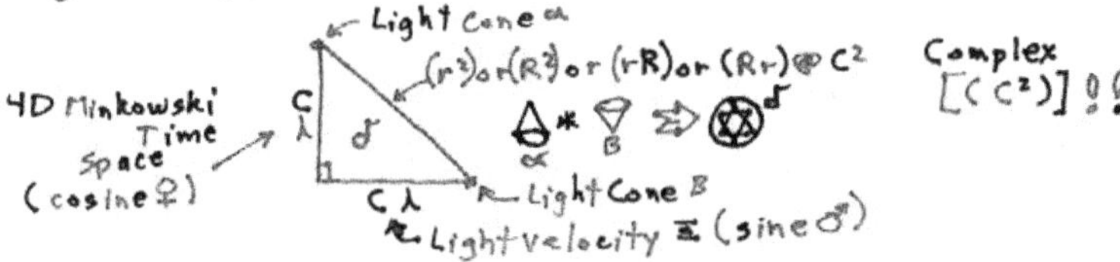

Time Travel is really Multi-Dimensional Superconnectivity

of IDEA & MEMORY using MUSIC & ART & COSMIC ENERGY in

combination [Hexi-Decimal Syncronization] with the Universal Circumference.

To build a time machine ∴ Develope the 'Theoretical Positron Store Ring'.

Set the NFBQ Nuclear Prism (Sonic Nucleon 'Noise' Feedback Loop

Quotient) above UNITY to PURITY. [1.T]!! Travel is

'Foreward thru Reversed Time' in 'Multidimensional MODE,'

From Cardinal Point α to Cardinal Point B. (NODE α $\Rightarrow$ NODE B)

The Fresnel Lens Array (Casmir) is the Time Machines 'Photon Clutch'.

C L 118

One is Unity @ Purity $1 (\equiv) 1$ is Identity

If $[1 (\equiv) 1]$ Then $[1 \rightleftharpoons \overline{1}]$ or $[1 \overset{\delta}{\circlearrowright} \overline{1}]$ ∴ Quantum Equivalence

With this in Mind; Idea & Memory flow into Reality.

At this Reality $\{\emptyset FF \nearrow 1 FF\}$ in the Cardinal Point $[\{\delta\delta\}]$

@ $\pi \int \pi$ or $\delta \circ m \delta \pi$

$\{$Genius is expressed as $\{\frac{Black}{White}$ as Gray$\} \Rightarrow \{\frac{Ultracolor}{Ultraclear}\}\}$

$\{$Brillance at Nodal Superconnectivity !!!

$[(1 \vdash 1) \circ \circ (1 \rightleftharpoons 1) \circ \circ (1 \overset{\delta}{\circlearrowright} 1) \circ \circ (1(\equiv)1)] \Rightarrow [\frac{(1 \cdot \overline{1})}{\delta}] \, [\emptyset \ell (\equiv) 1]^\delta$

$[\{[(*)] \overset{\pi \int \pi}{\Rightarrow} [\gg]^{(*)}\}] \circledast$

Cursor $\delta \circ m \delta \pi$ Ferndot

$[\emptyset FF \overset{*}{\Rightarrow} 1 FF]$

$\#[\gg]$ Superswitch

$\delta \quad 1 FF [\ll]$

space anti-
time space
time

$\emptyset FF$
$[*]$

πB ← $[$Photon Swirl$]$
πA (x, y, z, t)

$[\{$Butterfly $\rightleftharpoons$ Dragon$\}] \Rightarrow (1 \overset{\delta}{\circlearrowright} 1) \Rightarrow (1 \overset{*}{\circlearrowright} 1)$

Universal Circumference Cosmic Energy Superconnectivity. Thoughtform Amplification.

Cosmic Energy Blending at the Universal Circumference.

$\alpha \cdot \delta \, \pi \cdot$ $\pi \, \delta$ Feedback Loop [Noise]

$\emptyset FF \gg 1 FF$

$[$Complex Quantum Superswitch$]$

$[\emptyset \ell (\equiv) 1^\delta] \, \& \, [\overline{\emptyset \ell (\equiv) \overline{1}}^\delta]$

$[\frac{1 \cdot \overline{1}}{\delta} (\equiv) \emptyset] \circledast^\delta \, [\emptyset FF \rightleftharpoons 1 FF]$

$[\frac{1 \cdot \overline{1}}{\delta} \ggg \delta]$ | Casmir Photon Condenser

Thinking 'Singing Outside of the Box

$[$Identity
$(1 \vdash 1) \circ \circ (1 \rightleftharpoons 1) \circ \circ$ Photon Scalar Synchronization Equivalence $(1(\equiv)1) \circ \circ$ Quantum Superswitch $\pi A \, \pi B \, (1 \overset{\delta}{\circlearrowright} 1)]$
$\alpha \equiv B$ tangental Superconnectivity
πB

$[\frac{\emptyset \ell}{\emptyset} (\equiv) 1][\frac{\ell}{B}]^\delta$

$[\circ \circ] [\frac{1 \cdot \overline{1}}{\emptyset}]$ Superconnectivity Growing to Infinity

Cardinal Point

$[\emptyset \ell (\equiv) 1]^\delta \, [\{\emptyset FF \Rightarrow 1 FF\}]$

CL 119

In the case of 'one' (1) as 'Unity' ∴ Since (1) is the same as (↑)

Then Space/Time is the Inversion (1→) or 'one then Minus'.

Consider Quantum Logic ∴ $\left[\phi! (\equiv) 1^{\delta}\right]$ The Infinite Superconnectivity of all Inverse Zero is Unity

$\mathcal{O}(\equiv) M \lambda \lambda^{\delta}$

$\Sigma E (\equiv) M C^{2}$

$\left[\overline{\phi!} (\equiv) \overline{1}^{\delta}\right]$ The Inverse of Infinite Superconnectivity of all Absolute Zero is also Unity

The Rotation of the S to P plane in the Complex Equation

Consider the (g_{00}) Tensor @ the point of superconnectivity on the Universal Circumference of Now ↗ THEN

If $\left[(1)^{2}\right] (\equiv) 1$ Then $\left[1^{\delta} I (\equiv) 1\right] \notin \left[1.\phi (\equiv) 1\right]$

Quantum Harmonic Syncronization $\left\{\begin{array}{l}\notin \left[\phi! (\equiv) 1\right] \\ \notin \left[\phi! (\equiv) \overline{1}\right]\end{array}\right\}^{\delta}$

Matrix/Determinate $(g_{00})(g_{01})(g_{10})(g_{11})$

$(g_{00})(\equiv)\left[\begin{array}{c}\phi! (\equiv) 1^{\delta} \\ \notin \\ \overline{\phi!} (\equiv) \overline{1}^{\delta}\end{array}\right]$ $\therefore (g_{01}) \therefore (g_{10}) \Rrightarrow (g_{11})$

universal circumference IDEA Thought-form

Logic Superswitch $[\phi FF \Rrightarrow 1FF]$ Quantum Logic $[* \Rrightarrow \ll]$ [CHOICE]

π π^{α} π^{β}

Superswitch $\phi FF \Rrightarrow 1FF$ λ^{α} λ^{β}

γ Memory Node & Access

$c^{2} = \sqrt{\gamma^{2} * \ell^{2}}$

Identity @ (ϕ, δ) [TAU/THETA]

Pointeen Vector → Solition Wave → Universal Circumference

$c^{2} (\equiv) \sqrt{r^{2} R^{2}}^{c} (\equiv) \sqrt{r^{q}}^{c} \& \sqrt{R^{q}}^{c} (\equiv) \sqrt{!!}^{c}$

Unity & Purity

CL12ϕ

The Casmir Condenser & Amplifing Fresnel Lens Array
The Complex Equation ($\emptyset$ FF $\to$ 1 FF) takes Imagination °.°
Combines it with Creation Energy @ $\int_e \log^{e^{[B \to \emptyset \to B]}}$ and $\sigma \to \pi \sigma \int \pi$
and creates Reality as Mass along the Universal Circumference
as Solition Waves [δ] become Reality. Thoughtforms become
Real as Music & Art Syncronize with Growing [Multiplexed] Idea.

{Nuclear}
{Prism}

Photon Scalars
Wavefunction
(i)³
(i)³
Anchor

$\emptyset \sqrt{i}$ becomes δ [zero point derived]
Casmir Condenser is Final Focus Element
[(db, pq, λ)]
Fresnel Lens Array as Syncronizing Amp.
Energy Impulse from {Photon Scalar 1 π to}
{Photon Scalar 2 π}
Derivation of Cosmic Energy from the
Universal Circumference to Thoughtform Idea $\Sigma \to \delta \int \pi$

Augmentation of Reality. The Blending of Sound as Music & ART.
Brillance as Multidimensonal Superconnectivity

IDEA
SONG

Flash
Intigration
Quantum
Ressonance

How it's done°.° the clue is 'one as unity'. If 'one is one', then it is
unchanged @ 1 & Γ in all dimensions. Thus it can superswitch reality
from imagination. The Casmir Condenser and π Diamond (thru musical scintillation)
can create (creation energy) reality from idea. Think (Thoughtform) of
π^2 as [($\subset\pi$)($\pi\supset$)] with Slideruler
Then π (as each 'wing tip' or 'side of hypercube' or 'mirror')
Then π becomes the 'king post' or 'Lifting Surface of wing' or 'hypercube
[CL121] turning into Bubble' or 'Super switch mirror'.

Teleportation – Transmission of Energy thru Object
from time frame α to ℓ. $\oint \{ \begin{smallmatrix} \text{space} & \text{time} \\ \text{time} & \text{space} \end{smallmatrix} \}$

Given $E = MC^2$ $\qquad [\alpha \rightleftarrows \ell]^\sigma$

$\sum E_{(\Xi)} M \lambda \lambda \therefore \sum E_{(\Xi)} M \gamma \gamma$ If $[1 \overset{\frown}{(\Xi)} 1]$ Then $\sum M \gamma$ or $\sum \sigma$

$\qquad\qquad\qquad\qquad\qquad$ Identity $\qquad$ Object Solition Wave

Determinate / Matrix

$\begin{bmatrix} \alpha M & \{ \text{Rotated to } [(1)^3] \\ & \sigma \text{ Solition Wave on} \\ & \text{Universal Circumference} \\ & \text{Energy Impulse} \{ Q (\sigma) Q \} O \} \text{ Object Energy Matrix} \\ \ell W & \begin{bmatrix} R & r \\ r & R \end{bmatrix} \text{Plank Radius} \quad (db, pq, \lambda) \\ & \{ B \rightarrow \emptyset \rightarrow B \} \quad \text{electron within Nucleon} \end{bmatrix}$

$\{ \text{Rotated to } [(i)^3] \}$

Rotation to Higher Geometries $\gamma \sigma$

$\qquad [\ast \cdot \ast] \Rightarrow [\ll] \text{ Ferndot Flag}$
$\qquad \alpha \quad B$

Super switch @ $[\emptyset FF \rightleftarrows 1FF]$ $\quad \sum E_{(\Xi)} M \lambda^{(\ast \cdot \ast)} \lambda \Rightarrow \begin{smallmatrix} \text{Access to} \\ \text{Higher Dimensions} \end{smallmatrix}$

Solition Wave $\begin{bmatrix} R & r \\ r & R \end{bmatrix} [1 \overset{\frown}{(\Xi)} 1]$ $\quad \sum E_{(\Xi)} M \gamma^{[\ll]} \gamma \Rightarrow$
$\qquad\qquad\qquad\qquad\qquad\qquad\qquad \sum E_{(\Xi)} M \gamma \Rightarrow \sum E_{(\Xi)} M \sigma$

$\qquad$ Casmir

$\Psi_{\text{Wave function}}$ $\alpha \Rightarrow B @ C^2$

The Instanteous Split Second Infinitesimal for every
moment in time with the Nuclear Electron within the
Nucleon $(B \rightarrow \emptyset \rightarrow B)$ reaction and the Focus Determinate Matrix.

$$\sigma \oint_{\emptyset FF [\ast]}^{1FF[\ll]} \quad \sigma \int_{!}^{\alpha} \int_{\emptyset}^{B} \pi \pi \sigma \int_{\emptyset}^{\sigma} M \sigma \begin{bmatrix} \emptyset ! \\ \begin{bmatrix} R & r \\ r & R \end{bmatrix}^i \end{bmatrix} [1 \overset{\frown}{(\Xi)} 1]^{\sigma} \begin{bmatrix} 1.1 \\ ! \end{bmatrix}$$

$$[\emptyset (\Xi) 1]^\sigma$$

CL122

<u>Infinite Intrgration</u>   Superswitch Fulcrum thru Nucleon Prism

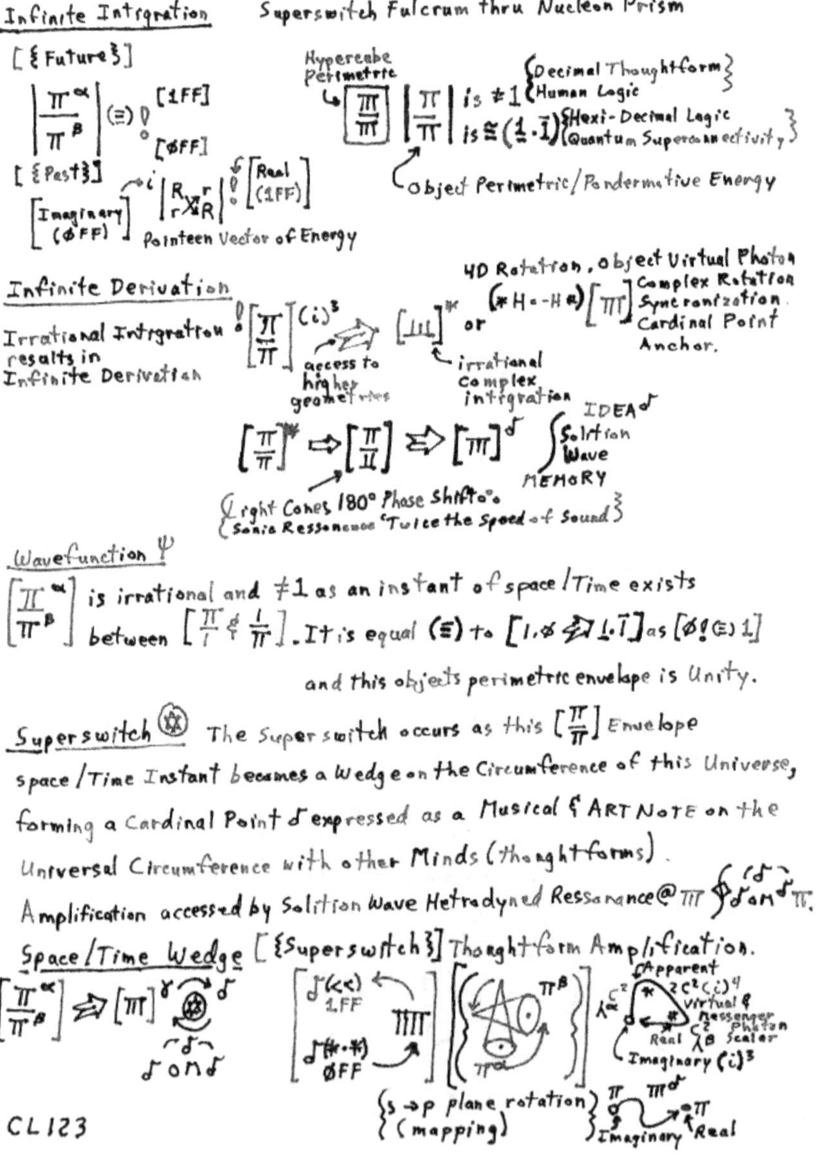

The 'Kingpost' of the Time Axis ($x \, c \, t$)

$$(x = \text{object} \quad c = \text{light} \quad t = \text{time})$$

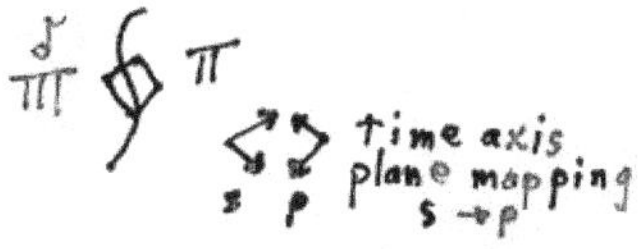

time axis
plane mapping
$s \to p$

$$\frac{d}{dx}\left[\int_{x_2}^{x^3}(e^{\sin t})\,dt\right] = \frac{d}{dx}\left[\int_{x^3}^{0}(e^{\sin t})\,dt + \int_{0}^{x^3}(e^{\sin t})\,dt\right]$$

$$= \frac{d}{dx}\left[\int_{x_2}^{0}(e^{\sin t})\,dt\right] + \frac{d}{dx}\left[\int_{0}^{x^3}(e^{\sin t})\,dt\right]$$

$$= \frac{-d}{dx}\left[\int_{0}^{x_2}(e^{\sin t})\,dt\right] + \frac{d}{dx}\left[\int_{0}^{x^3}(e^{\sin t})\,dt\right]$$

Since $\left[(1(\equiv)\!\mathit{L})\right]$ or $\left[(1(\ast)\!\mathit{L})\right](\neq)\left[\phi\right](\equiv)\left[\phi\!\int\right]$ or $E\, \xi\, M^{\sigma}$

$$\pi^{\alpha}(\ast)\pi^{\beta} \qquad \pi^{\alpha}\!\ll\pi^{\beta}$$

Final Derivation

$$G_{uv} = \frac{8\pi G}{C^4} Tuv$$

(Gravity Vector) / (Photon Bubble) (Time Vector)

(Surface Tensor in Space-Time
4th level Intigration (x y z t))

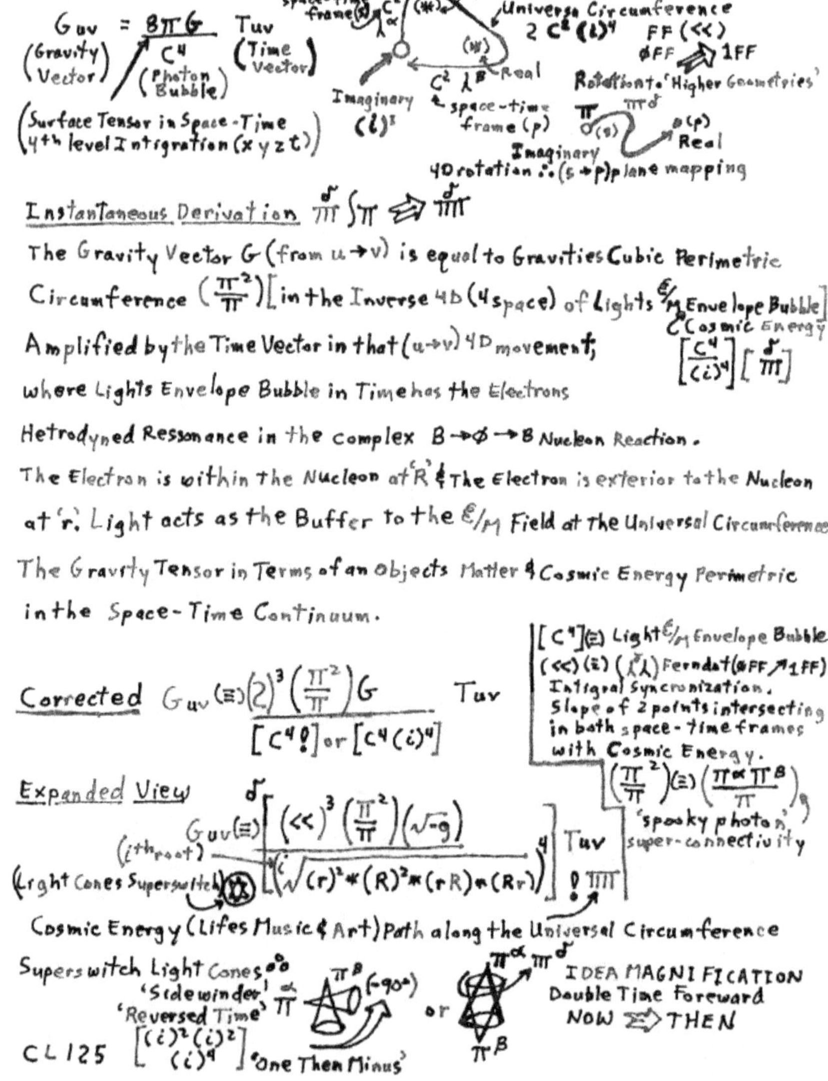

Instantaneous Derivation

The Gravity Vector G (from u → v) is equal to Gravities Cubic Perimetric
Circumference $\left(\frac{\pi^2}{\pi}\right)$ [in the Inverse 4D (4 space) of Lights E/M Envelope Bubble]
Amplified by the Time Vector in that (u → v) 4D movement; C Cosmic Energy $\left[\frac{C^4}{(i)^4}\right]\left[\frac{\delta}{\pi}\right]$
where Lights Envelope Bubble in Time has the Electrons
Hetrodyned Ressonance in the complex $B \to \phi \to B$ Nucleon Reaction.
The Electron is within the Nucleon at 'R' & The Electron is exterior to the Nucleon
at 'r'. Light acts as the Buffer to the E/M Field at The Universal Circumference.
The Gravity Tensor in Terms of an Objects Matter & Cosmic Energy Perimetric
in the Space-Time Continuum.

Corrected

$$G_{uv} \equiv \frac{(2)^3 \left(\frac{\pi^2}{\pi}\right) G}{[C^4 \phi] \text{ or } [C^4 (i)^4]} Tuv$$

[C⁴] (≡) Light E/M Envelope Bubble
(<<) (≡) $\left(\frac{\delta}{\lambda}\right)$ Ferndat (ΦFF ↗ 1FF)
Intigral Syncronization.
Slope of 2 points intersecting
in both space-time frames
with Cosmic Energy.
$\left(\frac{\pi^2}{\pi}\right) (\equiv) \left(\frac{\pi^\alpha \pi^B}{\pi}\right)$
'spooky photon'
super-connectivity

Expanded View

$$G_{uv} (\equiv) \frac{(<<)^3 \left(\frac{\pi^2}{\pi}\right)(\sqrt{-g})}{\left(i\sqrt{(r)^2 * (R)^2 * (rR) * (Rr)}\right)^4} Tuv$$

(i th root)
(Light Cones Superswitch)

Cosmic Energy (Lifes Music & Art) Path along the Universal Circumference

Superswitch Light Cones
'Sidewinder'
'Reversed Time' $\pi^B (-90°)$ or π^α π^δ

IDEA MAGNIFICATION
Double Time Foreward
NOW ⇒ THEN

CL 125 $\left[\frac{(i)^2 (i)^2}{(i)^4}\right]$ 'One Then Minus'

90

If $\left[1(=)1\right]^{(\mathcal{E}/m))}$ Then $\left[1(\equiv)I\right]^{(\mathcal{E}/m/G)}$ and $\left[1(*)I\right]^{4 \text{ axis } (x,y,z,t)}$

Identity — Unity — Purity

[Strong Force Electro Magnetisum] [Strong Force Weak Force Combined] [Strong Force & Weak Force Combined at the Universal Circumference]

IDEA 'one' — All 'ones Compliment' of that IDEA as MEMORY

Or $\left[\dfrac{1 \cdot \overline{1}}{\underset{\circ}{0}}\right] (\equiv) \left[\phi \, \underset{\circ}{0}\right]$ $\underset{\circ}{0}(\equiv)\left[(c^4)(i)^4\right]$ 'Photon Bubble in Envelope'

Quantum Equivolence

Centroid — Infinite (space-time) Flag or Cardinal Point

[(db, pq, λ)]

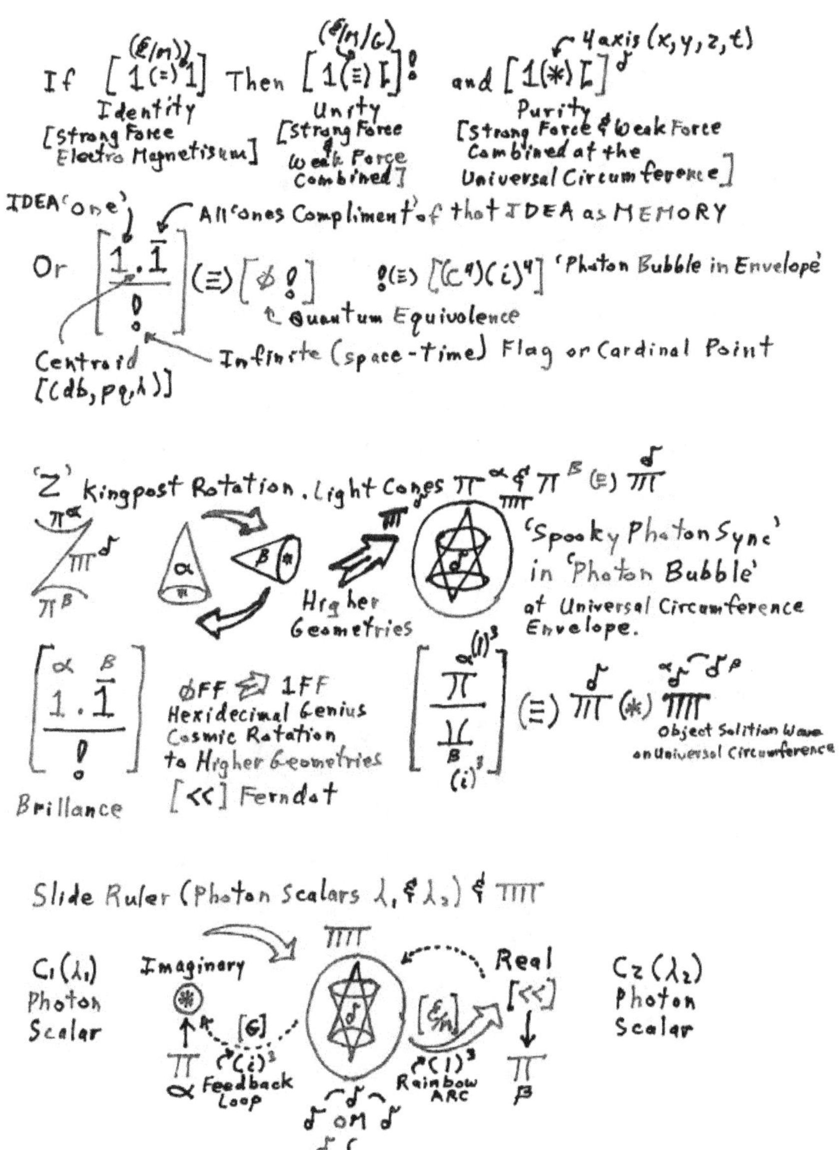

CL126

IDEA Amplification (Magnification)

The Individuals IDEA (as Thoughtform) becomes the
'Flame of LifesThought' as Light permiating the universe
for other minds to share and amplify. The IDEA
(as Thoughtform) causes that individuals DNA to
Blend with the Light of the Universe, Growing it as
Music and ART. The DNA itself crystalizes into TNA
becoming crystaliferous at the Universal Circumference
of NOW permitting other minds (as genius superconnectivity)
to Magnify in light as ART and Amplify in Music.
This becomes Song and Dance.

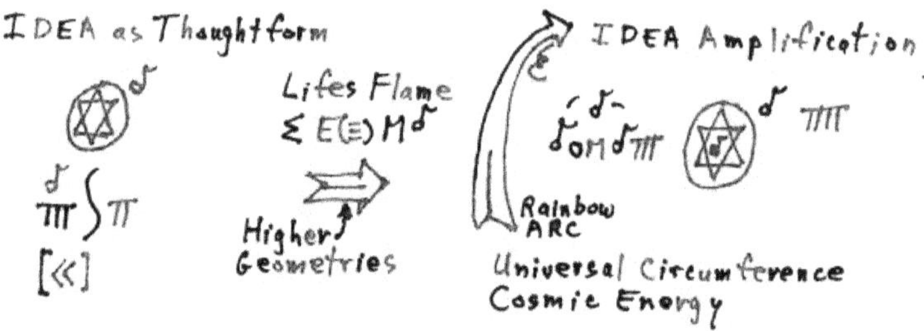

The Intigral Derivation of the Instant

space/time α ... anti-space/anti-time   Given two points α & B

Universal space-time curvature.   The space/time & anti-space/anti-time

space/time B ... anti-space/anti-time   instanteous interval between

IDEA & MEMORY ACCESS

The Formed Solution Wave @ the Universal Circumference

Blending the IDEA for sharing with other Minds

resulting in IDEA Magnification in ART & MUSIC.

The NODE or CARDINAL POINT as ∂! of each

IDEA Amplified as Light & Music

Twice The Speed of Sound. Acess to Higher Geometries ∂∂ @ $\frac{\pi \int \pi}{\partial \text{ and } \partial}$

Consider the 'Light Boom' of Cherenkov Radiation. A Photon Vacuum is

created. A Blending of Light as Electromagnetisum (the Strong Force) is

combined in the π Diamond as the 'Nuclear Prism' with Gravity (the Weak

Force). The Blending is the Energies of Two or More Light (Life) Forces

combining, resulting in an Ultralight FLASH that Instantly permiates

the Universe becoming a NODE or Cardinal Point in ressonance at

Ultraclear to all other Minds and IDEAS. This 'Light Boom'

originates at (.7C) and is amplified in the vacuum to

Ultralight Velocities becoming Instanteous and Multi-Dimensional!

CL 128

93

Blending of IDEA (IDEA Amplification)

(E/m) $(E/m/G)$ Cardino Point NODE   4 axis $(xyzt)$ Music & Art Solition Wave

If $[\,1(=)1\,]$ Then $[1(\equiv)I]^{0}$ and $[1(*)I]^{\sigma}$ object Cursor

Identity
{ Strong Force
  Electro Magnetisam }

Unity
{ Strong Force E/m
  Weak Force (Gravity) }

Purity
{ Strong Force & Weak Force Combined
  at the Universal Circumference }

Object Cursor ($*$) on the Universal Circumference σ σ Combined
Second Object Cursor Combined with the First $[(*\cdot*)]$ 00 Solition Wave Multiplexed Nodal Array

Dual Syncronized Object Cursors Multiplexed as Ferndot

$[(*\cdot*)]\,0\,!$ ⟹ $[(\ll)]^{\sigma}$ ⟹ ✡ ✡

Cursors   Ferndot Blend

IDEA Amplification

$\left[\, [1(*I)] ⟹ \begin{array}{l}\text{Thoughtform}\\ \text{Bubble}\end{array}\,\right]$ σ Cardinal Point

$\left[\begin{array}{l}\text{Hypercube}\\ \text{Memory NODE}\end{array}\right]$

Shared & Blended IDEA as Thoughtform on Universal Circumference

$\pi\pi(\equiv)\, \ll*\gg\, \sigma$ on σ Cardinal Point NODE

Purity $0\,0$ ← NODE
Unity   Ferndot Blend $\sigma\pi \overset{\alpha}{⊛} \pi^{\sigma}$ $\overset{\beta}{}$
$\left[\begin{array}{l}\text{Arbitrary Function}\\ (db, pq, \gamma)\text{ NODE}\end{array}\right]$ $\left[\begin{array}{l}\bar{Q}\\ Q\end{array}\right\}\circledast \left[(1-)\right]\right]^{\sigma}$

$\circledast$ Arbitrary Function, Note: Multidimensional $\frac{d}{dx}$ becomes $⊛\frac{d}{dx}$
$*$ as cursor becomes $\rightleftharpoons$ σ sine cos superswitch

$\circledast$ Arbitrary Function (Quantum Logic)
Minus '→' to $(i)^{4}$ $[(i)^{4}]$
'one Then Minus' to $[(i)^{2}(i)^{2}]$
4 D in 's' → 'p' $[(-1)(-1)]$
Plane Rotation
IDEA → REAL
IMAGINARY → IDEA

$[RAL/RAR]$ then
Rotate Accumulator Left $\rightleftharpoons$ Right

Corrected Memory Bubble

$\left[\begin{array}{l} Z \rightleftharpoons z^{2}+C\\ Z \overset{\rightarrow}{\underset{\sigma}{}} z^{2}⊛C \end{array}\right]^{\sigma}$

CL129

Corrected Split Second Integral with Arbitrary Function

The 'Kingpost' of the Time Axis $(x \, ct)$ $(x = \text{object} \quad c = \text{light} \quad t = \text{time})$

Time axis plane mapping $S \to p$ — Imaginary $\to$ Real

$$\frac{d}{dx}\left[\int_{x^2}^{x^3}(e^{\sin t})\, dt\right] = \frac{d}{dx}\left[\int_{x^2}^{0}(e^{\sin t})\, dt + \int_{0}^{x^3}(e^{\sin t})\, dt\right]$$

Cosmic Energy Correction

$$\sum (\equiv) MC^2/[(1)^3 \rightleftharpoons (i)^9] = \frac{d}{dx}\left[\int_{x^2}^{0}(e^{\sin t})\, dt\right] + \frac{d}{dx}\left[\int_{0}^{x^3}(e^{\sin t})\, dt\right]$$

Use the 'Arbitrary Function'

'One Then Minus'

$$= \frac{d}{dx}\left[\int_{0}^{x^2}(e^{\sin t})\, dt\right] + \frac{d}{dx}\left[\int_{0}^{x^3}(e^{\sin t})\, dt\right]$$

Superswitch Location
($\rightleftharpoons$ becomes $\longrightarrow\!\!\!\times\!\!\!\longrightarrow$)

Since $\left[(1(\equiv)I)\right]$ or $\left[(1^{(\#)}I\right](\not\equiv)[\emptyset](\equiv)[\emptyset!]$ or $E \lessgtr M^{\sigma}$

Superswitch @ Graviton Syncronization $\rightleftharpoons$ becomes $\longrightarrow\!\!\!\times$ Instantly

$\frac{\alpha}{\pi}(\#)\frac{\beta}{\pi}$ $\frac{\alpha}{\pi} \ll \frac{\beta}{\pi}$ α $\pi\pi$

Music & Art at the Universal Circumference

If $\left[\,I(\equiv)I\,\right]$ Then $\left[\,I(\rightleftharpoons)I\,\right]$ Then $\left[\,1(\equiv)I\,\right]_{\frac{0}{5}}^{0}$ and $\left[\,1^{\circledast}I\,\right]^{\sigma}$ Soliton Wave

(E/M) Identity   Quantum Identity   (E/M/G) NODE Cardinal Point
Strong Force
Electro Magnetisum   Superswitch   Unity Point   Strong Force (E/M)
&
Weak Force (G)

Gravity Waves are Multi-Dimensional [Photon Scalars]
Two types linking the Past & Future
{ One towards the Future $(1)^3$ }
{ One towards the Past $(i)^3$ }

CL 130

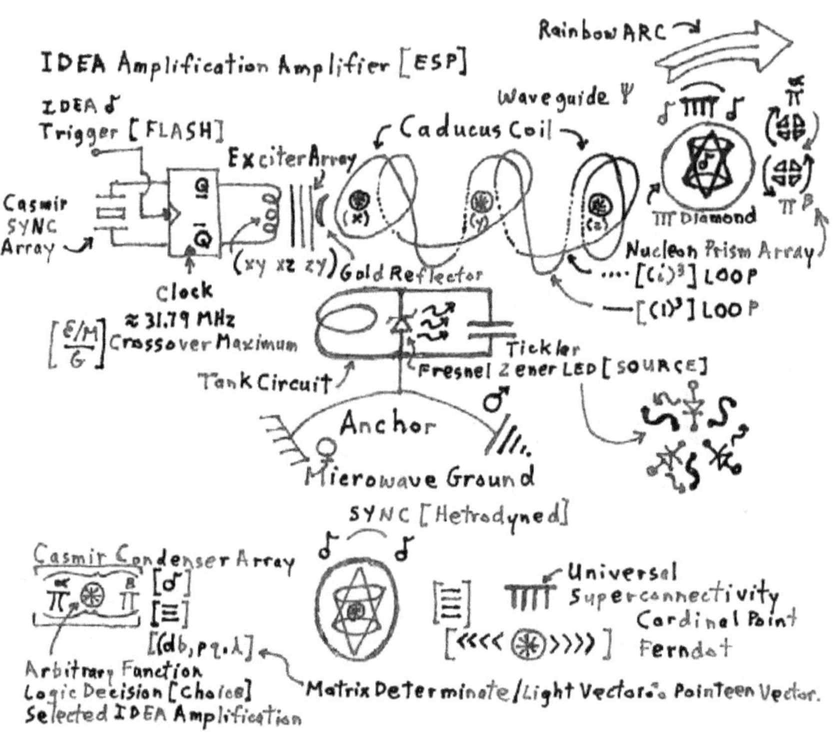

Rotation of Excited Caduceus Coil Array results in Photon Torque (+ or –) becoming an 'Arbitrary Function' ✷ ! !

Stacked Photonic Scalar Waves ($\propto$ & B) as 4D space-time Wedge.

CL 131

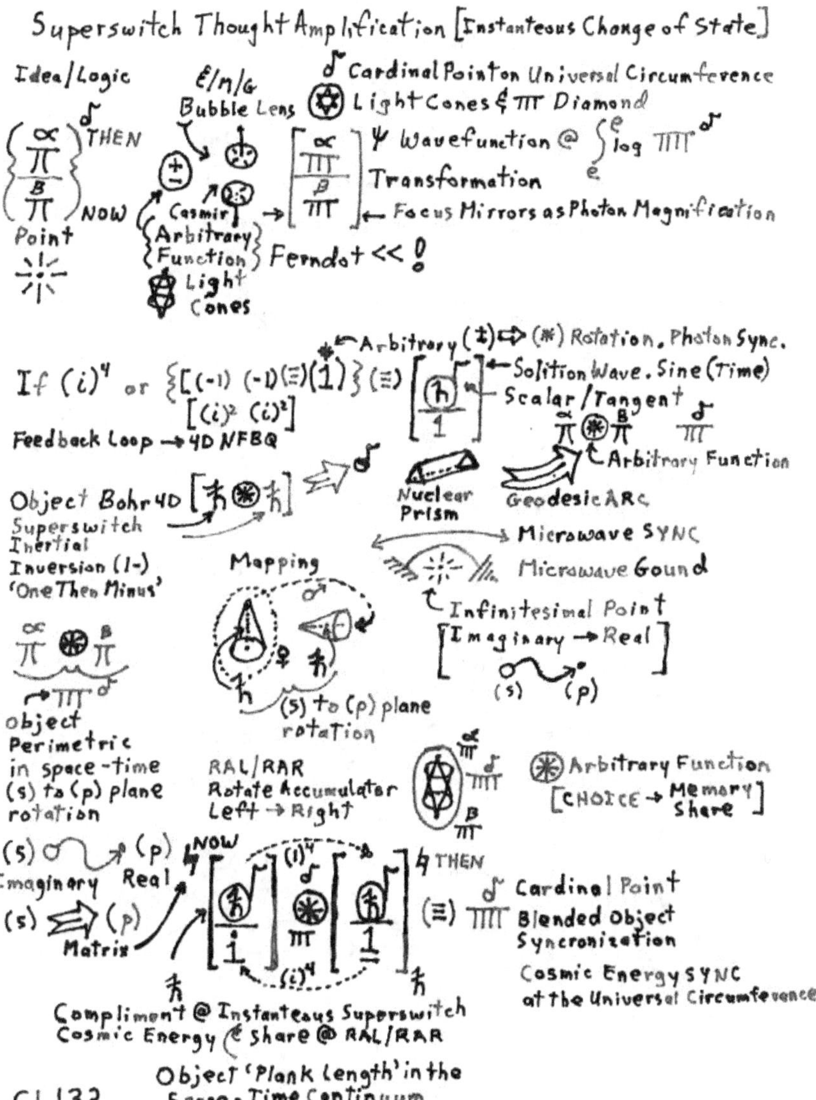

Superswitch Thought Amplification [Instanteous Change of State]
Idea/Logic
E/m/G
Bubble Lens
Cardinal Point on Universal Circumference
Light Cones & π Diamond
THEN
NOW
Casmir
Arbitrary Function
Light Cones
Point
Ψ Wavefunction @
Transformation
Focus Mirrors as Photon Magnification
Ferndot << !
If (i)⁴ or {[(-1)(-1)(Ξ)(1)}(Ξ)
[(i)²(i)²]
Feedback Loop → 4D NFBQ
Arbitrary (±) ⇨ (*) Rotation, Photon Sync.
Solition Wave. Sine (Time)
Scalar/Tangent
Arbitrary Function
Object Bohr 4D
Superswitch Inertial Inversion (1-) 'One Then Minus'
Nuclear Prism
Geodesic ARc
Microwave SYNC
Microwave Gound
Mapping
Infinitesimal Point
[Imaginary → Real]
(s)   (p)
object Perimetric in space-time (s) to (p) plane rotation
(s) to (p) plane rotation
RAL/RAR
Rotate Accumulator
Left → Right
(*) Arbitrary Function
[CHOICE → Memory Share]
(s)   (p)
Imaginary Real
(s) ⇨ (p)
Matrix
NOW
THEN
(Ξ)
Cardinal Point
Blended Object Syncronization
Cosmic Energy SYNC at the Universal Circumference
Compliment @ Instanteous Superswitch
Cosmic Energy & Share @ RAL/RAR
Object 'Plank Length' in the Space-Time Continuum
CL132

4D Superswitch   Teleportation. Rotation from the
'S' plane → 'p' plane. Foreward thru the Imaginary 4D space-time plane.
'Foreward thru "Reversed Time & Anti-Time".' 'One Then Minus'.
'2 and Thru'. 'Twice the Speed of Sound'. 'Overview & Looking Thru'.
IDEA Amplification. Memory Superconnectivity. IDEA Sharing.

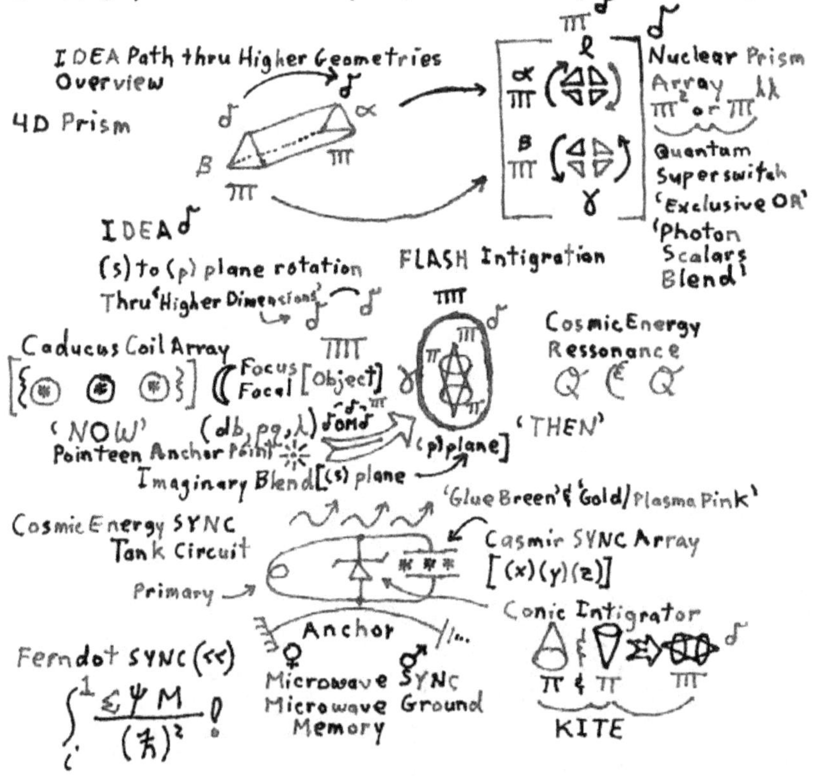

CL133

Superswitch 'Exclusive OR'. Quantum Superswitch with Flash access to Cosmic Energy and Multiuniversal access to the Universal Circumference. Thoughtform Magnification.

Thoughtform 'IDEA' and Memory Access Magnification.

Cosmic Energy Matrix

Superswitch Logic

Matrix Arbitrary Function (±) — Ferndot

Photon Scalar — Nuclear Prism — Universal Circumference

Thoughtform 'Logic/Idea' connects with the Universal Circumference as 'Music & Art'. Amplification (Multiplication) with additional Ideas then Memory. Music as {Memory/Superconnectivity}. (s) IDEA & MEMORY {plane rotation}.

'Time to Store & Multiplex.' 'Open' Photon 'Scalar' (p) 'closed' Cosmic Energy Access

'Flash' Array. Base 'Blend'. Higher Geometry Superswitch Array.

Photon Scalar Cascade {α & β} Sync [E/M/G] — IDEA Fusion & Magnification. (α & β)

FLASH @ Univesal Circumference

Fresnel Lens Nuclear Prism Array

Rotation to 'Higher Geometries'

Gravity Vector Swirl (v)(w)

[E/M/G] 'SYNC'

Arbitrary Function

[Focus / Focal]

(x) (y) (z)

Triple Diode Array 'Base' or 'Platform' [(x, y, z) Swirl] [(#•#)]

Real Memory (p)

'BLEND'

'CARRY'

Photon Scalar Solition Wave (s)

Imaginary IDEA

'SLOPE'(xct) Object Axis

[<<] EXCLUSIVE OR Superswitch Matrix

CL 134

CL 135

Genius ➡ Brillance

Rotation of Hypercube (Ultraclear Cosmic Energy Intigration)

IDEA ➡ REAL   Matter ⟹ Energy

$B\left(\frac{E}{M}\right)^B$ (i) $[(i)^i]^2$ $\dfrac{[(1)^3]^{\sigma}}{[\,\underline{i}\,]}$

E/M Bubble $(g)^2$ Imaginary Swirl ♭

$\begin{bmatrix} \text{Cosmic Energy} \\ \text{Sonic loop} \end{bmatrix}$ Gravity $\dfrac{}{g}$ $[(i)^2]$

Applied Energy Impulses along xct inertial axis $\begin{bmatrix} \text{Point} \\ \text{Slope} \end{bmatrix}$

Calculus Rotation using Complex Light Cones Interfacing

'Foreward Thru Negative [Inverse] Time @ $\frac{E/M}{G}$ ⁰

'S' ⟹ 'P' . 'S' to 'P', 'imaginary' & 'real' . Plane Rotation.

'S' plane   space $(1)^3$ ⟹ σ   time   real path

'P' plane   space   imaginary path $(i)^2$ time (i) (i) 'asymptotic'   Ferndot Syne (<<)

Light Cones Intersection (λ_1, λ_2) . Thought form Integration.

π & π Kite   Sidewinder   IDEA MATRIX   Calculus 'Discus' $\Delta xyzt$ Aribitrary Decision $[(\pm)]$

IDEA MAGNIFICATION.

Superswitch $[\{\emptyset FF ⟺ 1FF\}]$. Higher Multi-Dimension Path

Access to Higher Dimensions. Accelerated Dimensions & Complex Time

$C^2(\equiv) \sqrt{\gamma^2 * \ell^2} (\equiv) \gamma$ Memory NODE & Axis . $[\,1 (\equiv) \gamma\,]$ Cardinal Point ⁰

CL 136

The Wildfire Experiment.
Using the [{ØFF ⇄ ¹FF}] Statement to Superswitch
IDEA into Memory Then Energy into Mass forming Reality.
IDEA Magnification & Cosmic Superconnectivity become
Multi-Dimensional Amplification with access to Higher
Dimensions. Memory Intigration.
Multiplexed IDEA Array   Superconnectivity   Pondermotive Force
Event Augmentation.   Multiplication.   Superswitch.
Time Reversed Momentum [↑P] using the Arbitrary Function
(✳) @ Split Second Intigration becomes (✳)

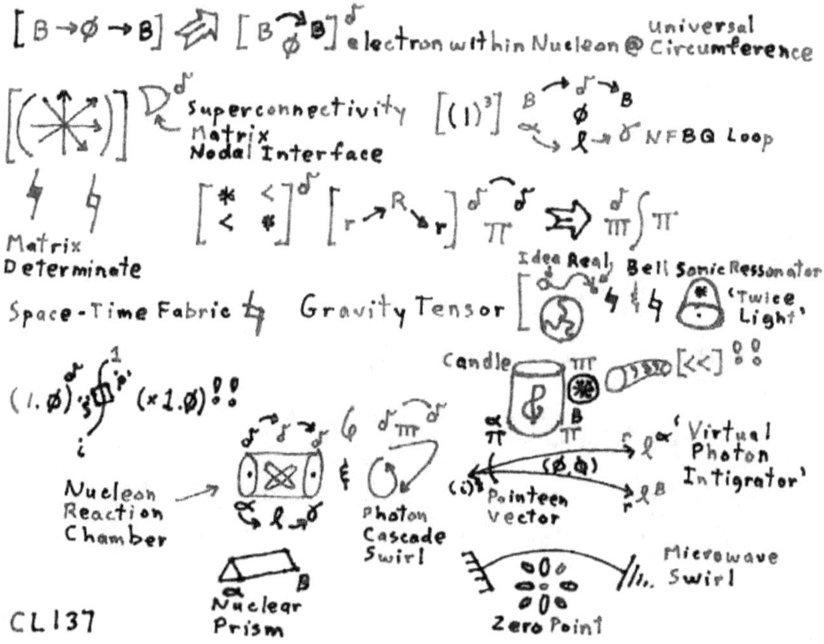

[B→Ø→B]   [B Ø B] electron within Nucleon @ Circumference universal

[(✳)]   Superconnectivity   [(1)³]   NFBQ Loop
        Matrix
        Nodal Interface

Matrix
Determinate

Space-Time Fabric   Gravity Tensor   Idea Real, Bell Sonic Ressonator 'Twice Light'

(1.Ø)   (×1.Ø)!!   Candle   [<<]   Virtual Photon Intigrator'

Nucleon   Photon   Pointeen   Microwave
Reaction   Cascade   Vector   Swirl
Chamber   Swirl

Nuclear   Zero Point
Prism

CL137

Slideruler @ [π↑ ✳ π↓]^{δπ} with 4D (Time) Augmentation
Slideruler (db, pq, λ).
E/M / Log
'Imaginary Expansion'
Yellow
Blue
'Strobe' 'Flash' Cursor
'Real' 'Instant'
π γ
π
δ
[Nuclear Prism]
π ℓ
Rotate Accumulator Left → Right (SKEW) RAL/RAR
Gravity / Log
Red
Green
δ OM δπ
IDEA
Imagination
Real Universal Circumference
Solution Wave (db, pq, λ)
Zero Point
Microwave Expansion
'Impenitrable Barrier'
Univesal Circumference
(s) (p)
λ₁
(i)³ Zero Point
λ₂
δ
(s) to (p) plane rotation
Rainbow ARC (Gray Scale)
δ
[(i)³]
Imaginary
Reversed Time
π (✳) π
Arbitrary Function 4D
Real [(1)³]
Time Foreward
E/M/G Bubble Augmentation
δ
space
supersolid
Curve (s → p)

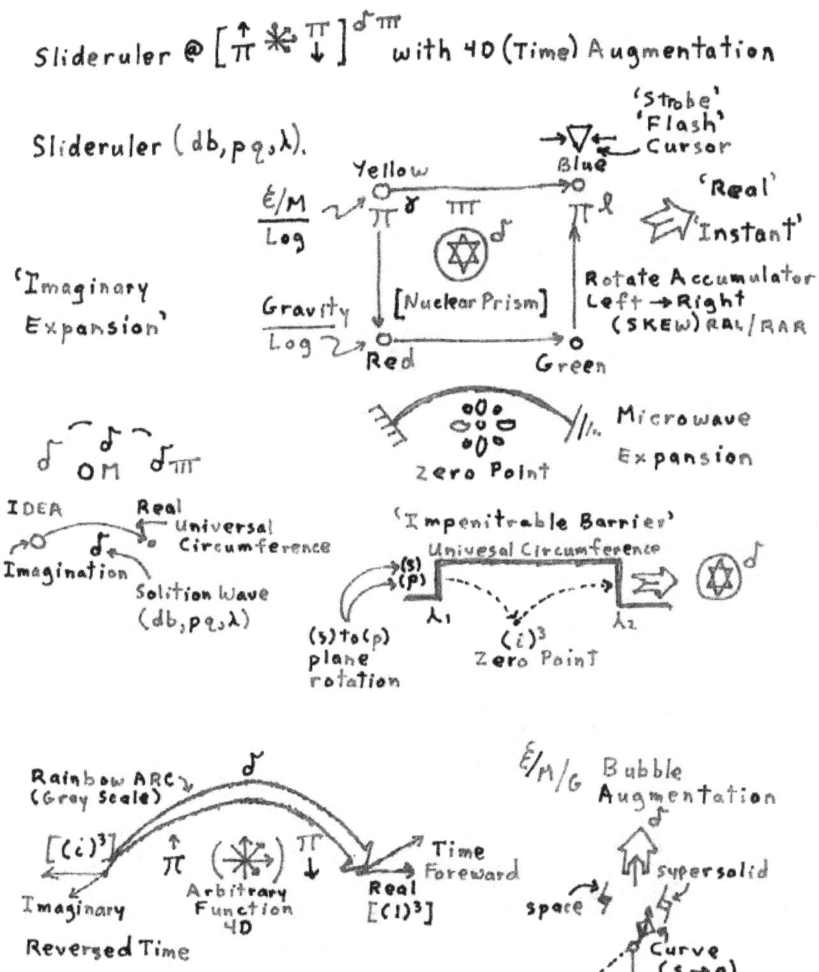

CL 138

Circuit Elements for Arbitrary Function Intigration.
Expansion into the Higher Dimensions of Space-Time.
The Time Axis as the only Instanteous Movement @ [B ∂ B]ᵈ.

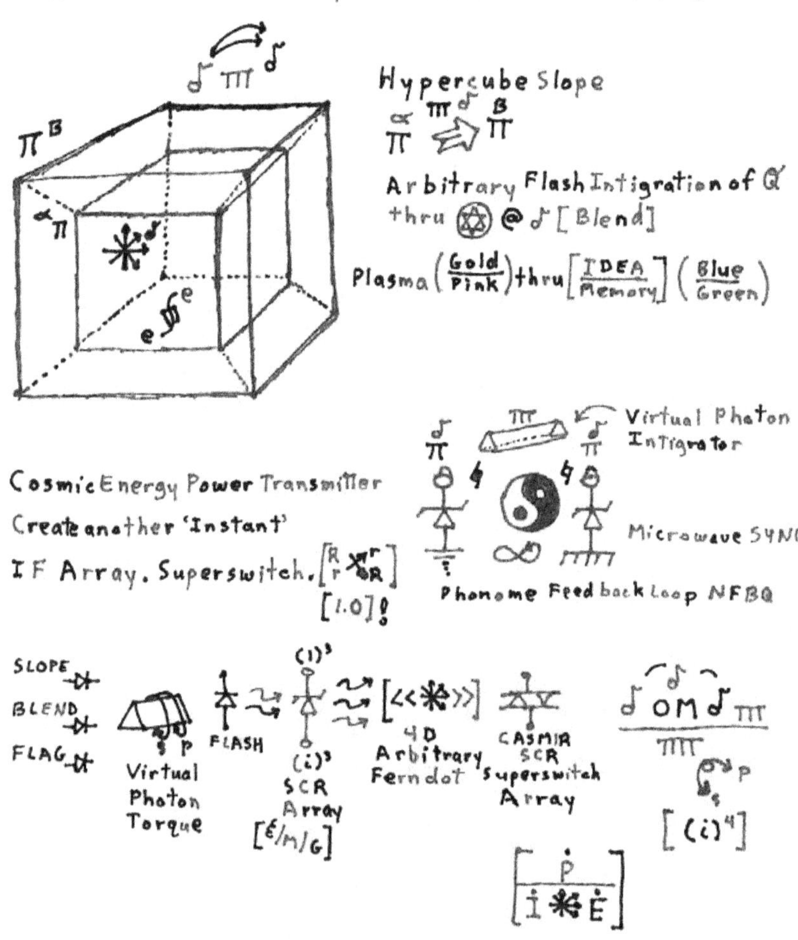

CL 139

Quantum Superswitch. Command Control of the Time Axis (xct).

105

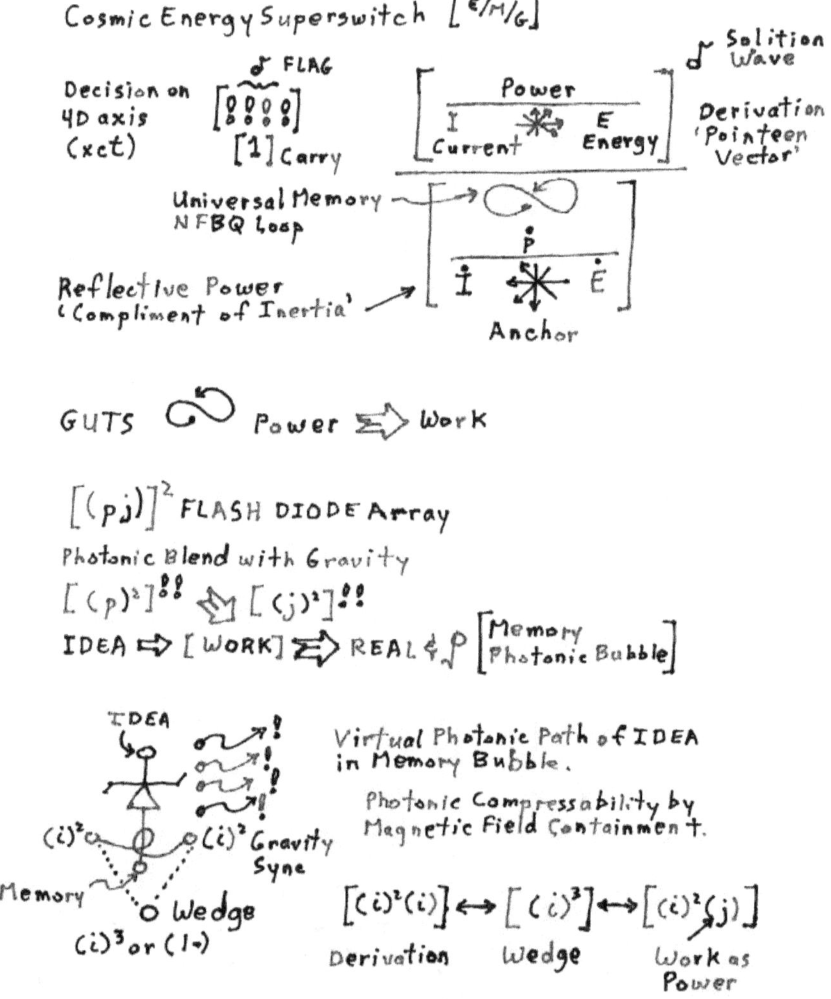

Cosmic Energy Superswitch [E/M/G]
Solition Wave
Decision on
4D axis
(xct)
FLAG
[0 0 0 0]
[1] Carry
Power
I   E
Current   Energy
Derivation
'Pointeen Vector'
Universal Memory
NFBQ Loop
P
I   E
Anchor
Reflective Power
'Compliment of Inertia'
GUTS   Power   Work
[(pj)]² FLASH DIODE Array
Photonic Blend with Gravity
[(p)²]   [(j)²]
IDEA   [WORK]   REAL & ∮ [Memory Photonic Bubble]
IDEA
Virtual Photonic Path of IDEA
in Memory Bubble.
Photonic Compressability by
Magnetic Field Containment.
(i)²   (i)² Gravity Sync
Memory
Wedge
(i)³ or (I∞)
[(i)²(i)]   [(i)³]   [(i)²(j)]
Derivation   Wedge   Work as Power
CL141

My Short Story

There is a source of Anti-Gravity on Route 5 just north of California called the Oregon Vortex. It was discovered when water flowed up hill near the cabin that collapsed into its anti-gravity well; built by the fellow that first discovered it. Well, I went to investigate it and decided that its' root cause was the "Ultra-Dense Toride" material that was in the meteor that fell there billions of years ago during Earth's formation. Probably star stuff formed in a super-nova; it still generated Scalar Waves that reversed gravity, and most likely space-time itself. I concluded I could use the US power grid, the Vortex as a theoretical position store, and GPS Wave Dynamics to generate a controlled Scalar Photon Stack (Field). It would be the first Hypercube. The story should start a little earlier. When I was about ten years old, my Uncle Vic and I were on my front porch when we both saw a ball lighting explosion; actually, it was the "reversed time end" of the explosion, a bright flash of light illuminating the entire area. The house across the street wasn't there, just the forest of trees from about fifty years earlier. The ball lighting's "theoretical position store" had touched the Commonwealth Edison power grid and grounded. The actual event occurred years later, when I was about sixteen, but the "reversed time flash" became an earlier event in time. Later, when I was about sixteen, I was in my backyard during a dark green sky's large thunderstorm. I saw a twister start to form, but instead of a tornado, it twisted a funny negative ninety-degree right backward angle, and its tip lit up as it touched the power plant's power grid, forming ball lightning and a theoretical position store. This explosion (at sixteen) traveled backward in time to me and Uncle Vic (at my age ten) and flashed a view of the far past where it ended. The view was the light of a forest of fifty-plus years ago when there were no houses here. It was a view that was as clear as a brightly lit day and lasted only for a flash of time. With this in mind, I eventually solved what I saw and wrote a book called the *Split Second Integral*, which the US Government would originally not let me publish when I applied for copyright until the "Freedom of Information Act" let me proceed. I had figured out how to generate a Hypercube using Earth's magnetic field, the Oregon Vortex theoretical position store, and Frensnel Lens (Nuclear Prism) Array. Incidentally, this is all near the Colorado area Tesla had selected for his laboratory and experiments (his Cosmic Energy Tickler.) An unexpected and very interesting side effect occurred, when the Hypercube that generated Cosmic Energy Fields caused DNA resonance to amplify thought into thoughtforms and DNA's higher crystal form TNA. Expanded consciousness, memory, and the resulting IDEA crystallization.

The first ESP related "Cosmic Energy Event" that I noticed occurred when I was seventeen years of age. I had a 125cc Suzuki street and dirt bike I rode all the time. One night, my friend Jon B. and I were going to visit a friend, and usually just rode without helmets etc. A small, one-fourth inch ball of red light appeared before my eyes as I passed by my helmet shelf, and it would not leave. I got the impression I was to wear my helmet; so I did. Late that night, we got into an accident, and the helmets took the brunt. I was tossed off the bike and rolled at least thirty-plus feet along the concrete road, and I'm sure I would have a crushed skull without the red ball's helmet warning. True! The helmets were a real mess. So was my leg. The second-most interesting result was my ability to switch my vision from normal to color to just black and white. No kidding! Just as a normal person can blink their eyes; I can turn my normal color vision to black and white only just by "blinking." Sometimes it has a darkish blue hue to it. I also sometimes think in a "Glue-Breen" (Blue Green normal) color and a "Goldish-Pink" color that is usually an orange color. It is called the "eighth colors" by scientists. I've been doing that since I was a kid.